Innovative Learning Geography in Europe

Innovative Learning Geography in Europe: New Challenges for the 21st Century

Edited by

Rafael de Miguel González and Karl Donert

CAMBRIDGE
SCHOLARS
PUBLISHING

Innovative Learning Geography in Europe: New Challenges for the 21st Century,
Edited by Rafael de Miguel González and Karl Donert

This book first published 2014

Cambridge Scholars Publishing

12 Back Chapman Street, Newcastle upon Tyne, NE6 2XX, UK

British Library Cataloguing in Publication Data
A catalogue record for this book is available from the British Library

EUROGEO

ISBN (10): 1-4438-5508-1, ISBN (13): 978-1-4438-5508-2

Table of Contents

LIST OF FIGURES

INTRODUCTION

Opportunities for developing innovative approaches in teaching and learning geography have been increasing very rapidly in recent years. This is in part because of the spread of new technologies that allow access to geographic information and geographic geo-media resources. Technological applications and user tools are readily available and this book examines aspects of their use in the classroom. These new tools that offer broad access to information and open data sources have revolutionised the way in which teachers of geography—in higher education, but also in primary and secondary education—can do their work with pupils and students. Education for Digital Earth, as conceived by Al Gore 20 years ago is now possible. The exclusive use of traditional approaches to the teaching of geography is no longer reasonable today.

The digital-earth.eu has identified many visualisation tools, Web sites, software developments, apps, didactic materials, Web-cartography and geographic geo-media. Many of these opportunities are already freely available on the Web accessed via Cloud-based services. The world of GIS, virtual globes and other forms of information representation and analysis offer immense pedagogical possibilities making the study of geography more attractive and effective for teachers and students. Geo-media allows the visualisation of information from different media sources and is concerned with digital content and its processing based on place, position and location. Cartographic communication has never been so easy to implement, thus twenty-first century school education needs to include geo-media into its daily workflow. Education for Digital Earth is critical if we are to make meaning of the world around us and learn how manage our environment and relate to others.

Innovative approaches to teaching and learning are needed to embrace study environments from local to global scales, for various reasons: empowerment in learning to excite and fascinate, technology providing potent tools and solutions to explain complex problems of the present world, enhancing learning processes, good classroom practice and building suitable training approaches. It is important to be aware that the media itself will not be the crucial factor for learning achievement, but the pedagogic approaches employed. The use of digital-earth tools can substantially enhance learning strategies and achievements when applied according to suitable, relevant and meaningful learning and teaching methods, such as active, student centred learning. In the case of geo-media, the development of spatial thinking and spatial citizenship as educational concepts leads those concerned to holistic learning experiences.

The European Commission-funded network initiative, digital-earth.eu, has promoted innovation and best practices in the implementation of geo-media as a digital learning environment for school learning and teaching. The network encourages the sharing of innovative practices. Some of them are described in this book, whose production was intended as a follow up to the European Conference held in the University of Zaragoza in 2012 about teaching and learning geography, with the support of the same network (www.digital-earth.eu) and EUROGEO, the European Association of Geographers.

The book is divided into two parts. The first comprises several chapters that analyse the main challenges facing geographical education. After firstly describing general issues addressed by this book and the importance of geospatial information in European school education, chapters two to five specify four major areas of innovation in geography education: curriculum, methodology, teacher training and geospatial technologies. The inclusion of these four main factors of pedagogical renewal is not baseless as they match the Special Interest Groups of the Comenius network digital-earth.eu. The second part of the book describes some on-going practices and illustrates different examples of the use of geoinformation in geographical education in different European countries and in various educational systems and contexts.

The opening chapter sets the scene, by defining the concept and movement of Digital Earth. It comments on the significance of geospatial technologies as big business today with many advances fuelling economic development, growth and planning. Europe has lagged behind the United States in recognizing the significance of this and though industry, science and technology are forging new horizons, Europe has been much slower to respond. The text reports on the activities of the digital-earth.eu networking initiative to raise awareness of the need for Digital Earth education in Europe and support teachers and trainers in dealing with this at Centres of Excellence throughout Europe (www.digital-earth-edu.net).

Teaching geography in schools is regulated by different curricula for school education. The second chapter offers a comparative study of the geographical curriculum featured in five European countries. The analysis carries reflections on common elements of the teaching of geography in Europe, but also shows differences in the inclusion (or exclusion) of content related to geospatial learning.

Spatial thinking/literacy in education should be a fixed component of school curricula and treated in a similar fashion to linguistic and mathematical thinking. It is an important part of everyday life. One of the best subjects to introduce spatial thinking in education is through geography. But, nevertheless, the introduction and use of GIS and geo-media in education has not yet been broadly acknowledged in most countries. Researchers have identified many reasons for this, mentioning among others the lack of education standards on the use of GIS in the school curriculum. The third chapter explains how the digital-earth.eu network has created and endorsed a benchmark with learning outcomes to be used throughout the primary and secondary education curriculum. To make it more practical this chapter suggests that a learning line in using GIS should be developed, with learning outcomes and activities with an increasing level of complexity.

As Geographic Information Systems (GIS) become more and more important in modern society, geography teachers and teacher trainers are increasingly interested in the possibilities of using GIS for teaching and learning. Teachers often have the feeling that they do not have the required knowledge to design and coach viable and effective geographic inquiry projects with GIS. The fourth chapter offers further insights into the nature of this issue. It describes some of the outcomes of a design study about the possibilities of using GIS in enquiry-based secondary geography education. The content suggests that teachers not only need to have sufficient GIS knowledge, but they must also have geographic and geographic-didactic knowledge in order to integrate GIS successfully in Inquiry-based geography education.

Teachers' and students' attitudes towards information and communications technologies are very important. The successful use of geo-media and geospatial tools depends partly on infrastructure, pedagogical use and leadership. The fifth chapter describes some of the leading resources and geo-tools for teachers, including GIS, and develops two case studies, one in a secondary school and the other in a higher education degree.

One of the reasons for school success in Finland has been the open minded implementation of educational innovations: this is replicated in Finnish geographical education. Thus the second part of the book (Chapter Six) begins with a review of the Finnish targets of acquiring and using high quality geographical information. A particular goal has been to make diverse use of information technology to interpret processes and to present this information. These skills should be developed throughout geography education. In upper secondary school optional courses on regional studies using GIS data and applications are also offered. Incorporating GIS in the Finnish school curriculum has had multiple benefits. It enhances spatial perception skills, improves understanding of the environment, and promotes sustainable way of living, as well as develops skills required in working life and enhances digital literacy skills. The text shows how a web-based GIS learning environment PaikkaOppi has resolved these issues and offered a high quality and easily accessible tool to support the learning of spatial information in Finnish schools. The study showed that Finnish teachers have a high appreciation for the free, web-based learning environment that allows them and their students to engage openly with GIS. They also think it is important to teach with GIS, not about GIS.

The seventh chapter describes a prototype course that uses GIS to teach Albanian middle school students about urban sustainability subjects. The course is based on free online and off-line data and software. The topics address both Albanian and global subjects, while the software includes the ArcGIS Online Map Services and Google Earth. The course, and related data and software, and the products created by the students are freely available online. Pedagogical principles that guided the design of the course include: teaching with GIS rather than about GIS, integration of technology across many media forms, integration of concepts across disciplines, connection of students' personal experience to the larger world, a mixed-age classroom rather than a single age classroom, balance of students' role as consumers of knowledge versus that as creators of it, and a networked classroom structure versus a hierarchical one.

The aim of Chapter Eight is to compare the GIS education in the geography curricula of two countries, Turkey and Portugal, how the approaches to GIS education vary, to what extent the curricula deal with GIS education and in what way. It examines the main similarities and differences. It is important to ask such questions within this research in order to understand how and why GIS has been placed in the curricula.

Chapter Nine explores an educational participatory approach built around a geo-referenced e-tool created to identify packages of integrated e-services for tourists. It asks whether digital cartography can help to synthesise feelings, ideas, values, and land-use projects. The project aimed to create a network of local actors and students by using a bottom-up approach and create direct connections with the landscape. The results are a map of the city of Genoa constructed in a cooperative/collaborative way that shows the efficiency of the geo-referenced website as a facilitation tool for the young people involved in the project.

Another national case is shown in Chapter Ten where the applications of GIS in Greek secondary education are described, as well as the research, pedagogical materials and results of introducing GIS and geospatial technologies in school classrooms. Moreover, it suggests cartography can be utilised by students to approach concepts such as numeration, measurement, patterns, relationships, functions, data, probability etc. The map, however, as an absolute metric spatial tool, based on "syntactic" rules, illustrates space in an abstract form. As a result, a map in apprehending space requires two different, but simultaneous, approaches: the map as a cognitive object and the map as an object of spatial knowledge. Chapter Eleven thus presents a framework with examples of how to reinforce cognition in space, through instruction with maps, demonstrating that digital earth technologies and spatial thinking are integrative and cross-curricular with specific applications for other subjects of primary and secondary education.

The book concludes with two chapters (12 and 13) explaining different experiences of the use of geo-media tools and resources for the study of several issues about Spanish geography, such as landscapes, and Spanish Geographic Institute map resources.

Building the capability to introduce geospatial information, tools and technologies in education requires commitment from leading educators by challenging tradition, developing and building ideas and creating innovative materials for classroom use. It also requires that the geospatial industry encourages and supports these leaders, allowing them to sustain their actions and efforts. Decision makers also need to be made aware of the state-of-the-art and advised in terms of how geo-media should be incorporated in programmes and curricula. Civil society organisations and citizens must also become more involved. At this moment in time there is no forum to bring these actors together. The digital-earth.eu initiative has started to break down some of the barriers and focus on connecting these stakeholders.

Rafael de Miguel and Karl Donert, Editors

PART ONE:

GENERAL ISSUES

CHAPTER ONE

BUILDING CAPACITY FOR DIGITAL EARTH EDUCATION IN EUROPE

KARL DONERT

Introduction

In 1992, former U.S. Vice President Al Gore presented a farsighted Digital Earth concept, whereby detailed geospatial information could be accessed from any place, at any time, by anyone (Gore, 1992). The subsequent scientific and technological movement has made this vision a reality today. Based on a US Department of Labor study, Gewin (2004), writing in the scientific publication Nature, proposed that geo-technology (with related spatial thinking skills) would become one of three most significant technological advances for economic development in the next decade. Since then, in the United States there has been a strong lobby for geospatial education, resulting in Congress's acknowledging the significance of the National Academies Press publication "Learning to Think Spatially" (National Research Council, 2006). This has transformed the US research and education technology agenda and, as a result, the National Science Foundation (2011) recently awarded significant grants to geospatial education research. In Europe most developments have been haphazard, small scale, and without backing from political stakeholders.

Digital Earth and European school education

The Digital Earth vision expressed by Al Gore linked groups of scientists interested in cooperative studies of the planet and its resources (Gore, 1998). The initiative directed technology and research actions towards solutions for sustainable development. Since then, advances in digital earth technologies have created a profound revolution in science and technology. Strong societal connections have been established as a result the rise of the Geo-web and use of social media.

The acquisition and use of geospatial information, combined with developments in computing and communications has made near real-time information about the earth available to billions of people. The Digital Earth concept has become a reality and the results play an increasingly important role in addressing the social, economic, cultural, scientific, and technological challenges affecting the way we understand the earth. Digital Earth allows scientists to focus their attention on many of the important challenges faced by Europe today, such as economic efficiency, resource depletion, sustainable energy, natural hazards, food and water supplies, environmental degradation, population migration and smart cities. Politicians are beginning to realise the immense opportunities Digital Earth offers in everyday decision-making processes.

Access to information enables citizen participatory processes (Turnhout et al., 2012). Recent developments of geographic geo-media can be used to bridge the gap between citizens, Digital Earth technologies and real-world problems by socially connecting them through geographic location. Geo-media therefore has the capacity to create powerful learning opportunities that can empower students and result in flexible, individualised learning based on critical thinking and approaches that can explore complex interdisciplinary issues. Despite this potential, European education, for instance in science, history, geography, media studies and ICT, has so far, by and large ignored the opportunities afforded by these Digital Earth developments. This is despite the fact that geo-technology has become a significant employer and geoinformation and geo-media have become almost ubiquitous commodities accessible from mobile, tablet and laptop.

In school education, geo-media can help students to construct spatial concepts and promote a meaningful understanding of our world through problem solving, experimentation, project work and the communication of findings to others (Gryl, Jekel and Donert, 2010). The visual elements offered by geo-media are essential for enquiry, exploration and communication. However, here are only small pockets of intense activity (Kerski, 2008) and geo-media education in Europe has generally lagged behind (Donert, 2010), especially concerning its implementation in schools and teacher training.

Creating a digital-earth.eu European network

Research has confirmed that in Europe little or no attention has been paid to the significance of emerging geospatial technologies in schools (Milson et al., 2012; Donert, 2010; Gaudet and Annulis, 2003). A few pilot projects have been funded to create teaching resources in several languages. A number of face-to-face and online training courses have been successfully delivered to relatively small numbers of teachers and educators. It is apparent that large-scale, ministerial-initiated implementation in education has been generally lacking, indicating that European education has generally been unable to keep pace with technological and societal changes taking place. Awareness of the significance of these technologies among these stakeholder groups remains low, despite the recent initiatives on ICT, jobs and skills encouraged through the Digital Agenda for Europe (http://ec.europa.eu/digital-agenda/).

During 2009, in response to geospatial developments and the absence of centralised initiatives, an Austrian Centre for geo-media education (digital:earth:at) was created centred in Salzburg and linking a number of Austrian organisations who were working with schools and teachers. The goal was to share resources, tools and innovative ideas to increase the use of geo-media with Austrian pupils and teachers. Its successful launch and implementation resulted in the development of a proposal for a networking initiative, called digital-earth.eu, connecting stakeholders across Europe. The result was a proposal consisting of 49 partner organisations from more than 20 countries. Funding was obtained from the European Commission Lifelong Learning Programme (http://eacea.ec.europa.eu/llp/) for them to collaborate together for three years (2010-2013) under the Comenius Programme for schools and teacher education.

At the heart of this development was the creation of an infrastructure centred on a European Centre of Excellence, based at the Austrian Centre of Excellence (Lindner-Fally, 2009). The aim was to build a Community of Practice based on individuals and organisations that could support teachers and schools in different parts of Europe, connecting people working in national and regional contexts (Jekel et al., 2008). The digital-earth.eu Comenius network sought to raise awareness by educators of the many innovative 'geospatial' developments taking place and reflect on their implication and potential impact on school education systems. Another purpose of the networking project was to influence policy makers who had already begun to connect European issues involving social and environmental developments to citizens. A series of lobbying activities were undertaken, predominantly by the European Association of

Geographers (EUROGEO), through digital-earth.eu. This led to political engagement with the EC 'Digital Agenda for Europe', 'New Skills New Jobs' and 'EyeonEarth' initiatives. Dissemination activities promoted the incorporation of 'education for digital earth' into regional, national and European educational agenda. Politicians and decision makers at different scales were addressed and informed.

Digital-earth.eu would allow those involved in the network to connect with one another, share ideas and information, communicate future visions, and develop an informed Community of Practice (CoP) (Li et al., 2009). The CoP was to be based on a network of expert Centres for geo-media across Europe. An evaluation of proposed members was undertaken through a peer review process and accredited by the European Centre and the European Association of Geographers (EUROGEO). These expert Centres should form multipliers by working with many teachers and trainers in their own situations and contexts. They were also able to offer advice and guidance to Ministries of Education and decision makers at national, regional and local levels. This process offers increased visibility to organisations that are doing outstanding work; it encourages and supports innovation in learning and teaching approaches and rewards quality. At the time of writing this chapter, sixteen Centres in 14 European countries have been established, and two others are going through the review process.

The activities of the digital-earth.eu project address a broad range of issues. These include teacher training standards, professional development and geo-media competences. They consider issues of data availability following the results of the EU INSPIRE initiative and the tools available for educators to use.

At the core of the digital-earth.eu network have been four thematic special interest groups affording opportunities for collaboration in specialised areas. In the project proposal these were defined as:

1. Data, Tools and Technologies
2. Learning and teaching environments
3. Teacher Education and Training
4. Curriculum developments

A needs analysis of network partners confirmed the importance of these themes and confirmed that while technical advances have extended the Digital Earth vision in scientific terms (Gore, 1998; Foresman, 2008; Goodchild, 2008), in education their uses were still mostly restricted to a few users within schools and teacher training. There has been an explosion in the number of geospatial Web 2.0 tools available for teachers to use with their students, yet digital earth technologies were not widely described in national curricula. Most European Ministries of Education and even the European Commissioners for Education and the Digital Agenda remained largely unaware of their existence.

These groups reviewed the state of the art and contributed to an online catalogue of materials, courses, publications, links and best practice scenarios as well as producing a series of research papers, publications and guidance materials. Dissemination through social media and a series of electronic newsletters sought to keep those involved up-to-date with developments and resources.

The digital-earth.eu project team recognised that it was almost impossible for most teachers to keep pace with the plethora of technologies at their disposal. The Data, Tools and Technologies group examined many of these new resources and opportunities and created a database and geo-services to promote their availability in school and teacher training contexts. These tools and technologies included social media, media content like RSS feeds, blogs and video clips, open apps freely available to download for mobile devices, mashup interfaces (Al-Khudhairy and Delilah, 2010) that allow interactive on-the-fly mapping, sophisticated visualisations and geo-collaborative activities developed via distributed Cloud-based, Web GIS (Alexander, 2006).

The group explored some educational perspectives of the outcomes of the European INSPIRE initiative and examined the possible impacts for teaching in schools and in teacher education. They then reviewed data availability, standards and interoperability and addressed property rights from a school perspective, producing publications to inform teachers and teacher educators. This resulted in a series of recommendations for action. A report was produced which explored issues related to freedom of information developments across Europe encouraged by the INSPIRE Directive and the Digital Agenda. It considered issues like copyright, Intellectual Property and quality issues concerning data and information in different European countries relating to schools and teachers. Volunteered geographic information (Goodchild 2007) and crowdsourcing (Howe, 2008) were examined as interesting alternatives to traditional information sources from mapping agencies and companies. An online searchable catalogue of resources was created which provides an infrastructure through which resources, data, information and teaching materials can be shared.

Digital earth technologies can be used in education as tools to encourage enquiry and problem-based learning and enhance critical thinking and geo-communication (Kriz et al., 2013), construct personalised teaching materials, and assist students' self-expression (Beak et al., 2008). The second working group looked at learning and teaching concerns that were connected with the use of geo-media in schools. There are many different aspects that can play a determining role in successful learning. Their focus developed on learning environments created and used in schools and classrooms. They examined student-centred learning approaches, using geo-media in transmissive, dialogic, constructivist and co-constructive ways (Mishra & Koehler, 2006) where teachers are encouraged to create guided enquiry approaches in their classrooms (Powell, 1999). The role of digital storytelling opportunities was considered highly significant, encouraged by Web 2.0 tools and communications technologies (Levine, 2010).

The group reported on key competences in the use of geo-media, examining the concept of geo-media literacy and made recommendations for the inclusion of spatial competences as key competences for lifelong learning. They then undertook a review of learning and teaching approaches and provided practical guidance for teachers and teacher educators. A publication (in press) will introduce different learning and teaching approaches to teaching with geo-media and geoinformation by examining comparative approaches and including exemplars, highlighting best practice. This publication will be connected to a conference dealing with aspects of e-learning, geo-media and spatial citizenship in teacher education and schools.

It was confirmed that Digital Earth technologies offer opportunities for meaningful, deep learning experiences in and beyond schools. It contributes to teaching and learning by supporting exploration and experimentation; it improves motivation and learner engagement and offers the learners more responsibility and control through individual and group communication (Kolacny 1969). The research undertaken confirmed that European education must focus on spatial thinking, so that learners will understand spatial patterns, linkages, and relationships (Bednarz et al. 2008).

The third working group addressed the complexity of pre- and in-service teacher education. Kerski (2008) discussed the important role teacher's play in using key technologies to prepare students to be tomorrow's decision makers, where they are able to tackle local, regional, and global 21st century issues. The group recognised that teachers remain key components to an effective use of computers and geo-technologies in the educational system (Zhao et al., 2001). They established developing positive attitudes towards using technology in education is essential and confirmed research by Teo et al., (2007) that demonstrated how teacher attitudes towards new technologies are a major predictor of successful uses.

The report produced by the group reviewed the state of teacher training and geo-media and makes recommendations for benchmarking. It confirmed that support must be offered to help teachers develop positive attitudes toward computers (Kadijevich and Haapasalo, 2008). To achieve this, the group created the European Centre for teaching and training in geo-media and produced a business plan that would establish an infrastructure of Centres of Excellence across Europe to support teachers and trainers at grassroots level. The group also looked at quality enhancement issues in training and the formulation of an agreed terminology and a benchmark statement for geo-media. Research was undertaken to report on teacher accreditation across Europe (Lindner-Fally et al., 2012) and the opportunities for certification and accreditation in geoinformation. A booklet for teacher training will be produced to offer a checklist and guidance on incorporating geo-media/GI for those training teachers. It will deal with in-service training and continuing professional development of teachers.

Educational technology plays an important role in moving from teacher-centred learning activities to student-centred learning activities. It is therefore essential to have trained teachers competent in using and managing educational technology (Smarkola, 2008). The working group confirmed that the main remaining challenge was to convince education management stakeholders across Europe that the adoption of Digital Earth tools in their classrooms and training sessions both enhances the way they work as well as improves their effectiveness as teachers.

The final special interest group examined the curriculum opportunities for using geo-media and geoinformation in schools. This is concerned with the situation that, as most teachers have a strong sense of subject identity, they are predominantly influenced by disciplinary concerns. However, as Kerski (2008) suggests, today's main challenges lie with transforming the general structure of our educational systems to meet the needs of society. Geo-media applications tend to provide cross-curricular opportunities challenging traditional curriculum development. This group is developing a series of case studies of best practice, gathered through the Centres of Excellence and from earlier projects and initiatives to illustrate how to open access to the use of geo-media to all pupils. This publication will provide examples in main curriculum areas, including mathematics, languages, science, history, economics, business studies, marketing and geography. It will illustrate some techniques used to engage pupils and some of the outcomes from the classroom. The group also produced resources and guidance that target curriculum creators and programme developers, to advise and guide those involved in developing curricula, creating courses and lessons using geo-media. It also examined professional connections and links between schools and enterprise.

The management of change in education will become very significant if we are to embrace the new geo-media environments that encourage personalised learning. Their adoption, adaptation and integration in education cannot currently keep pace with the rapid growth of Cloud-based apps and geo-browsers offering access to state-of-the-art geo-technologies. Projects like digital-earth.eu are essential for the future of the industry if education is to match the rapidly increasing demands for a geospatial workforce. In future, capacity building of a professional profile and school-to-career developments will be needed if geospatial industry development is to be continued and the increasing demand for geo-media professionals can be met. These aspects remain to be addressed in the future.

Widespread network dissemination has sought to reach as many relevant organisations as possible, including teacher associations, Ministries, academies and other relevant institutions in 'hard-to-reach' situations. The goal has been to raise the profile of learning with digital geo-media, encourage innovative practices and reward organisations and individuals displaying 'excellence'.

Conclusions

Originally education was fundamental to the original Digital Earth concept, as Joseph Kerski (2008) commented:

"The Beijing Declaration on the Digital Earth recommended that Digital Earth 'be promoted by scientific, educational and technological communities, industry, governments, as well as regional and international organisations' (Xu and Chen, 1999). The declaration emphasised 'understanding the oneness of the Earth and its relevant phenomena.' It called for "adequate investments and strong support in 'scientific research and development, education and training'". (Kerski J., 2008)

However, educational perspectives of Digital Earth have not received as much attention as other areas. The digital-earth.eu project is a direct extension of the original Digital Earth initiative. The European Centre was invited to become a member of the International Society for Digital Earth in 2013 (http:// www.digitalearth-isde.org/). The project has raised awareness of the importance of geo-technologies and geo-media and has stimulated further innovative developments in the uses of geo-media in schools and education across Europe, for example through the Spatial Citizenship project (http://www.spatialcitizenship.org). The digital-earth.eu project has attracted considerable interest from researchers who are seeking to make advances in curriculum, learning and teaching approaches, teacher training and awareness of useful tools and technologies.

The digital-earth.eu network project was founded to raise awareness of geospatial education and inform politicians and Ministries of the significance of digital earth tools and technologies. It has been developed to connect organisations involved in geospatial education, so they can share practice, provide advice and guidance on the use of geographic (geo-media) to others and be a place for innovative future thinking and new initiatives.

The growing shortage of a geospatial workforce in Europe (Schultz et al., 2013), the significance of open data, freedom of information and the EU INSPIRE Directive suggests that Digital Earth education and training developments are urgently needed as part of educational structures like the European Qualifications Framework (European Commission, 2008). European policy makers have to be made much more aware of geospatial concepts (Marsh et al., 2007; Strobl, 2008) and then actively encouraged by stakeholders to respond to them in policy terms. This work is ongoing and needs to continue through the accredited Centres of Excellence and in developing a "Digital Earth education for all".

References

Donert K (Ed.) (2010), *Using Geoinformation in European Geography education*, Vol. IX, International Geographic Union-Home of Geography, Rome, 145pp

European Commission (2008), *The European Qualifications Framework for Lifelong Learning (EQF).* Luxembourg: Office for Official Publications of the European Communities

Gaudet, C., & Annulis, H. (2003), Building the Geospatial Workforce, *URISA Journal*, 15 (1), 21-30

Gewin, V. (2004), Mapping opportunities. *Nature 427*: 376-377. 22 January

Gore, A., (1992), *Earth in the Balance: Ecology and the Human Spirit.* Boston: Houghton Mifflin.

Gryl I, Jekel T and Donert K (2010), Spatial Citizenship, In Jekel T, Donert K and Koller A (Eds.) *Learning with GeoInformation V*, Berlin, Wichman Verlag, p. 2-12.

Jekel, T., A. Koller, and K. Donert, Eds. (2011), Learning with *GeoInformation VI*. Heidelberg: Wichmann Verlag

Kerski, J. (2008), The role of GIS in Digital Earth education, *International Journal of Digital Earth*, 1(4): 326-346

Kriz, K., Cartwright, W., and Kinberger, M. (2013). *Understanding different geographies* (pp. 1-6). Springer Berlin Heidelberg.

Li, L. C., Grimshaw, J. M., Nielsen, C., Judd, M., Coyte, P. C., & Graham, I. D. (2009). Evolution of Wenger's concept of community of practice. *Implementation Science*, 4(1), 11, http://www.implementationscience.com/content/4/1/11, accessed 3/6/2013.

Lindner-Fally M (2009), Digital:earth:at – Centre for Teaching and Learning Geography and Geoinformatics, *Proc. HERODOT Conference*, Ayvalik, Turkey, http://www.herodot.net/conferences/Ayvalik/papers/geotech12.pdf, accessed 10/11/2012

Lindner-Fally M, Herlander Mira H, Silva DV, Carvoeiras LM, Lambrinos N, de Lazaro y Torres M-L, Schmeinck D, Zwartjes L and Donert K (2012), *Teacher Education and Training and geo-media in Europe* http://213.235.245.69/fileadmin/deeu_documents/4_1_report_teachereducation_final.pdf, accessed 10/11/2012

Marsh, M., R. Golledge, and S.E. Battersby (2007), Geospatial Concept Understanding and Recognition in G6 College Students: A Preliminary argument for Minimal GIS, *Annals of the Association of American Geographers* 97(4): 696-712

Milson, A. Demirci, A, Kerski, J. (Eds.) (2012) *International Perspectives on Teaching and Learning with GIS in Secondary Schools*, New York, Springer

National Research Council (2006): *Learning to think spatially. GIS as a Decision-Support System in the K-12 curriculum*. National Academies Press, Washington DC

National Science Foundation (2011), *NSF Geography and Spatial Sciences (GSS) Program*, http://www.nsf.gov/funding/pgm_summ.jsp?pims_id=503621, accessed 5/10/2012

Schulze, U., Kanwischer, D., and Reudenbach, C. (2013). Essential competences for GIS learning in higher education: a synthesis of international curricular documents in the GIS&T domain. *Journal of Geography in Higher Education*, 37(2), 257-275.

Strobl, J. (2008), Digital Earth Brainware, In J. Schiewe, and U. Michel. (Eds.) *Geoinformatics paves the Highway to Digital Earth* (gi-reports@igf) University of Osnabrück. Osnabrück, 134-138.

Turnhout, E., Van Bommel, S., and Aarts, N. (2010). How participation creates citizens: participatory governance as performative practice. *Ecology and Society*, 15(4), 26, http://dare.uva.nl/document/227945, accessed 2/7/2013.

CHAPTER TWO

INNOVATIVE LEARNING APPROACHES TO SECONDARY SCHOOL GEOGRAPHY IN EUROPE: NEW CHALLENGES IN THE CURRICULUM

RAFAEL DE MIGUEL GONZÁLEZ

Teaching and learning Geography in Europe: Spatial competences

An important reference that should be taken into account in any process of curricular revision of the subject of geography is the International Charter on Geographical Education, published by the Commission for Geographical Education of the International Geographical Union, in 1992. From the point of view of the implementation of geographical education, a series of matters stand out in the Charter, such as the fact that geography is both a common core and an independent subject, organised in coherent syllabuses in compulsory school years for all the students, developed with a similar timetable to the other compulsory subjects, with specialised teachers, etc. All this brings out a first distinction between Mediterranean countries such as Spain, France and Italy, where geography is studied as a joint subject along with history (and social studies), and the remaining European countries, mostly those with the Anglo-Saxon scope, where geography is an autonomous school discipline.

Notwithstanding these determining factors, there are other issues in the Charter that have epistemological and methodological relevance to the very same process of teaching geography. Thus, the text highlights the students' need to acquire knowledge, skills and attitudes. In other words, the Charter contains conceptual contents, as well as procedural and attitudinal, specific of the teaching of geography. Although this document was written a decade and a half before the European document on eight key competences for lifelong learning (2006), it significantly reflects some of these competences, i.e. social competence and environmental competence. Moreover, geography fosters the acquisition of communicative, intellectual, practical and social skills, as well as the acquisition of linguistic, mathematical and visual skills.

The British subcommittee of the Commission of Geographical Education considered, in 2007, that the above mentioned document should be revised owing to the advances made in both the discipline of geography and its teaching, and that it should include a number of relevant matters such as globalisation, the moral and ethical aspects of geography (those that refer to the notions of power, influence and control), the issues related to poverty, equality and social justice, referred to both people and places, the education for sustainable development, the impact of new technologies of geographic information, the qualitative research in geographical education, interdisciplinary issues and space from a constructivist point of view, etc.

That same year (2007) the Commission for Geographical Education itself drew up a new declaration, complementary to that of 1992, devoted specifically to geographical education for sustainable development. In this new text, several specific geographical competences in order to improve the teaching and learning about sustainable development are listed (once again expressed in terms of knowledge, skills and values) and which must be supplemented with cross curricular skills, i.e. skills aimed at responsibility and action, and also with the necessary skills for a long life learning process and for citizenship participation, which might resemble to social and civic competences.

Moving from international documents to European ones on geographical education, one must set out from the fact that owing to the principle of subsidiarity jurisdiction regarding education is a national concern (or in the cases of Germany and Spain, the concern of their regional governments). Nevertheless, the European Union can take steps in order to support, promote or complement the decisions made by the member states themselves. In this way, there are several initiatives to undertake a review of education, a fact that allows a bigger connection between this and the improvement of competitiveness and employment. Just to mention a few: the implementation of the European Higher Education Area (EHEA), the mobility, networking and innovate actions of the Lifelong Learning Programme, the Comenius, Erasmus, Leonardo programmes etc., the setting up of the European Qualifications Framework or the definition of eight key competencies for lifelong learning, the

Strategic Framework for European cooperation in education and training (ET 2020), the Copenhagen Process, etc.

Recommendation 2006/962/EC of the European Parliament and of the Council of 18 December 2006 on key competencies for lifelong learning expresses eight major competencies, although geography as a school subject is not referred to and spatial thinking is mentioned only as a mathematical way of thinking inside mathematical competence. Nevertheless, competencies in science and technology, as well as those concerning social and civic competences are related to the teaching of geography in most European countries' curricula.

Besides digital competences, the European Digital Agenda (2010) is an essential reference when reviewing the future geography school curriculum. The Digital Agenda set up the acquisition of digital competences as a key and prime action in the regulation of the Social European Fund, as well as identifying the competences of professional people and those of the users within the framework of ICT as far as the above-mentioned European Qualifications Framework. This has some evident implications in the case of geographical education, especially in the use of geographical information in the classrooms (Donert, 2010) as a tool for "spatially enabled learning", the main axis of which is social geo-communication (Vogler et al. 2012).

In turn, the use of geoinformation as a didactic resource in teaching geography involves the acquisition of two further skills: spatial citizenship (Gryl, Jekel and Donert, 2010) and spatial thinking (NRC, 2006). The first competence includes three specific skills: the handling of techniques and methodology of spatial information, assessment and reflection on spatial representations and communication and citizens' involvement in spatial representations. The second competence recaptures once more the educational challenge of spatial thinking, no longer from the mathematical side as it was reflected in the European document on key competencies from 2006, but from the side of teaching geography as a way to develop spatial intelligence (Kerski, 2003), quoting the terminology of multiple intelligences by Gardner. This issue has been particularly taken into account by the National Board of Education in Finland at the time of setting up the curriculum for secondary education, and, especially the syllabus of school geography from a constructivist approach (Houtsonen, 2006): in this educational system geoinformation is not only important for the development of skills related to spatial citizenship, sustainable development, cultural identity or new technologies: its use in the classroom fosters the development of logical thinking from geo-referenced data, and, consequently the ability to solve problems of spatial nature. In future, this is likely to be required as a basic skill in many professions. As a matter of fact, competence in spatial thinking is known as competence for understanding space and covers three key aspects in the curriculum of geography: spatial visualisation, spatial orienteering and building of knowledge based in spatial relations.

The European Comenius network digital-earth.eu confirms that the use of new technologies of geographical information are essential for the students to become competent in those issues and claims that it is basic for geo-media to be included in the school curriculum and teacher training, even much more than they appear in the current curriculum (De Miguel, 2011), because of five arguments: i) they are likely to be necessary in future jobs; ii) they improve the students' conditions of employability; iii) it contributes to social competence and active citizenship; iv) the technologies of geographical information are an essential element in our daily lives; and v) it promotes teachers' innovation in the teaching of geography to understand and explain the present-day world. The importance of these matters has led the European Geographers Association (EUROGEO) to produce a manifesto[1] on behalf of strengthening geospatial abilities in the education and literacy in geoinformation, both for geography teachers and students at primary and secondary education.

In short, the curricular reviews of geography are a pending challenge for European education. Apart from the epistemological evolution of geography as a science, innovation in geographical education faces important prospects due to the following four huge challenges:

- The new European guidance on key competencies, where spatial competencies (spatial thinking, spatial citizenship) holds its own autonomy, this is closely related to the teaching of geography in schools, as such competences are being worked into other school subjects. And vice-versa, geographical education promotes the acquisition of other school competences, such as social, civic, digital, etc.
- The reduction of traditional, descriptive approaches and their replacement by learning methodologies that are much more active and based on inductive processes such as inquiry-based learning, case studies, problem-based learning, spatial assessment, etc.
- The teaching-learning processes necessary to address big social, political, economic and cultural issues in the global agenda, present and future, which have an effect on space, i.e., globalisation,

---

[1] A *manifesto for Europe: building geospatial capacity*, which is available on http://www.digital-earth.eu/documents0.html

sustainability and environment, climate change, urban development and housing, energy management, population pressure and migration, food and water supplies, transportation etc.

- The progressive usage of GIS and geo-localised digital devices that make use of geographical and cartographical tools, easily accessible on the web as well as the acquisition of skills and working procedures proper of the geographical work and of the geographical methods. And, consequently, the needs of the geospatial industry.

The incorporation of these four elements in national geography curricula in secondary education is unequal: there are some countries, which have been able to innovate and update their curriculum to include the issues previously cited, while there are others who maintain a traditional, compulsory curriculum, that does not allow any kind of curricular autonomy to the secondary schools and teachers, and impose conditions on the teaching methods and approaches to geography in the classroom. Showing these differences is the main aim of this chapter since there has been very little research into comparative geography curricula across Europe. A pan-European survey was conducted as part of the HERODOT project, but it was into geography higher education and not at the secondary level (Donert, 2007). Curic et al. (2007) compare the curriculum in eleven European countries, but most Mediterranean countries (Spain, France, Italy…) were not included. The proceedings of the Symposium on Curriculum making in geography organised by the IGU Commission on Geographical Education provided contributions from several countries, but without an evaluative synthesis (Whewell, et al., 2011). And even the report of digital-earth.eu on Curriculum Opportunities for Geoinformation in Europe (Donert, Parkinson and Lindner-Fally, 2010) only dealt with project partners.

The absence of a study that analyses the components of educational innovation (including spatial competencies and geoinformation) in the national geography curricula justifies the remainder of this chapter, as it focuses on England, Germany, France and Finland and reviews the importance of the forthcoming geography curriculum reform in Spain.

The Curriculum of Geography in England

In secondary schools in England geography is taught in three stages: at Key Stage 3, which lasts from 12 to 14, most pupils study geography as a national curriculum subject. At Key Stage 4, ages 15-16, geography is an optional subject. However, at the end of this stage there is a final exam of all secondary education (the well known GCSE exam), which includes the geography contents of Key Stages 3 and 4. Geography is one of the most popular subjects at Key Stage 5, with over 32,000 and nearly 46,000 taking an exam at either A Level or AS Level respectively.

In Key Stages 3 and 4, the most relevant features are: place, space, scale, interdependence, physical and human processes, environmental interaction and sustainable development, and cultural understanding and diversity.

The British geography curriculum in secondary education includes a series of basic procedural content, such as geographical research at school, collection of geographical information and spatial data, fieldwork, use of geoinformation, mastering of cartography, acquisition of specific vocabulary, the understanding of spatial processes by means of the study of cases and the methodology of problem-based learning, etc. All these conclude in attitudinal content to understand and value the environmental changes and those caused by sustainable development

This curriculum is much more open and synthetic than other countries like France and Spain as regards the enumeration of content, but much more practical and richer as far as resources used to master certain skills and methods to study geography are concerned. This enables the wider use of active and inductive methodologies, i.e. inquiry-based learning. And there is consequently much less use of master classes and memory-based learning of geographical facts. This curriculum thus makes possible the application of constructivism to the learning of geography, and greater motivation of the students.

This is demonstrated by the assessment criteria in use, and the specifications for the evaluation of the external GCSE test from the main examination boards: about fifty per cent of the mark corresponds to content related to geographical skills such as graphical, cartographical and statistical approaches, but also to methods of understanding and researching in geography content, which contribute to the acquisition of the competencies of spatial thinking and spatial citizenship, as it appears in the draft of the new National Curriculum for first teaching in schools from September 2014. These proposals include (at Key Stage 3) two learning aims related to the use of geo-information, firstly that pupils should taught to "use Geo-graphical Information Systems (GIS) to view, analyse and interpret places and data" and secondly to "use fieldwork to collect, analyse and draw conclusions from geographical data, using multiple sources of increasingly complex information".

At Key Stage 5, for pupils aged 16-18, this approach is strengthened, since the curriculum is not organised by conceptual content (which is usual in the Spanish or French curriculum) but in blocks of understanding, interpretation and application of geographical knowledge. For example, in the A Level specification (from AQA, the largest examination board) the subject of geography is divided into four big thematic blocks: physical and human geography, geographical skills, geographical topics about the contemporary world and fieldwork, geographical assessment and research. In the first year (A1) the first block deals with conceptual content frequently found in physical and human geography. This amounts to 70 per cent of the final mark. The second deals with geographical, statistical, graphical, cartographical, computer techniques, etc. (which amount to the remaining 30 per cent). The second year (A2) leads to the A level final examinations, the third and fourth blocks are incorporated with the following percentages: 35 per cent (first block), 15 per cent (second), 30 per cent (third) and 20 per cent (fourth). This distribution is justified by a series of specific educational objectives for the teaching of geography, directly related to the spatial thinking and spatial citizenship enhancement.

The Curriculum of Geography in Germany

In 2007, the German Geographical Society published a report on the educational standards of geography for secondary education. This report was delivered to the *Kultusministerkonferenz* (composed of Ministers of Education from the respective federal states (*Länder*), with a "positive response" from each of them. After an introduction that argues about the contribution of geography to compulsory education, six areas of competences specific to geographical education are detailed as central competences (Figure 2.1). The first three competencies (K, SO, M) match spatial thinking, whereas the others (C, E, A) align with aspects of spatial citizenship.

These competences has been closely translated into the curriculum for each of the regions of Germany, for since the competencies are defined by all federal states, the programme of study may differ from one state to another. As an example, in the city-state of Berlin the curriculum for geography in compulsory secondary education (7-8-9-10 *Klasse*) has notably reflected the recommendations by the Geographical Society when it determines the five specific competences of the subject, which are complementary to the acquisition of geographical knowledge and skills: spatial orientation, spatial analysis, spatial perception, spatial assessment and spatial awareness. Complementing these competencies, the assessment criteria are set for both stage 7-8 and stage 9-10.

Area of competence	Central competences
Subject-specific knowledge (K)	Ability to understand spaces at different scales as physical and human geographical systems and to analyse the interrelations between man and environment.
Spatial orientation (SO)	Ability to orientate oneself in space (topographical orientation, map-reading competence, orientation in real spaces and reflection upon spatial perceptions).
Gathering information/methods (M)	Ability to collect and evaluate geographically/geoscientifically relevant information in real space and in media, as well as to describe the steps in the gathering of information in geography.
Communication (C)	Ability to understand geographical information, to express and present it and to discuss it appropriatly with others.
Evaluation (E)	Ability to evaluate spatial information and problems, information in the media and geographical insights in terms of specific criteria and in the context of existing values.
Action (A)	Ability and willingness to act in accordance with natural and social conditions in various fields of action.

Figure 2.1. Key Competencies for Geographical Education.
(German Geographical Society, 2007)

For their part, the content has a bias towards regional geography: a list of six topics in stage 7-8 (eastern Europe and northern and central Asia; Monsoon Asia; Far East; Middle East; sub-Saharan Africa; Maghreb) and four in stage 9-10 (America; bioclimatic diversity in the planet; global sustainability and climate change; Germany and Europe). On the one hand, this curriculum is very detailed and allows the teacher very little choice or flexibility, but on the other hand, it is very explicit in the formulation of active methodologies. In fact, the curriculum analysed above is completed with a list of case studies of the remaining regions and topics (threatened cultures, Oceania, Seas and Oceans, Polar regions), and with a list of didactical recommendations for the assessment of geography, for the achievement of 'geographical workshops': following an inquiry-based learning methodology, the students choose the topics, following a cross-curricular approach and encouraging the use of new technologies and GIS.

In the instance of upper secondary education, the curriculum regulates a first introductory phase of physical geography (atmosphere and climate change, hydrosphere and water cycle, lithosphere and soils) and one of human geography in terms of spatial challenges (demographic, food and water supply, energy, environmental and sustainable development). After that, there are four blocks of content, distributed into four semesters (two each year), orientated towards an approach to descriptive geography of the big regions in the planet, with a special emphasis on the incorporation of Germany in the world system and global economy: settlements and spatial planning, Europe, developing countries and world economy regions (USA, Asia-Pacific). In short, geography assures the acquisition of the six quoted competencies, and the understanding of the complexity of the contemporary world and the political, economical and social relationships, which explain the spatial transition of the big regions of the Earth. At the same time, this curriculum also encourages the use of "getting, editing and evaluating" geographic information through the new technologies such as the Internet, GIS and virtual globes for spatial citizenship.

The Geography Curriculum in France

The Decree of 15th July 2008 has set the teaching programme for geography, history and civic education for the 6th, 5th, 4th and 3rd years of *collège* (11 to 14 years old). It is surprising that in this text the term curriculum is not used, although a constructivist approach to the planning of the teaching is clearly reflected in several sections: each block of content is thus divided into knowledge, skills and abilities (as assessment criteria). In the overall block of geographical content in compulsory secondary education, a specific use of active methodologies can be distinguished: inquiry-based learning, case studies to introduce the students to geographical research and to illustrate the most theoretical knowledge to local and regional spaces. In this approach, the analysis and presentation of geoinformation in the case studies is considered to be a main learning tool, as shown by advanced geographical techniques.

However, there is still a strong influence of the regional paradigm and the regional school, neglecting the educational contributions by other geographical schools. Likewise, the almost complete absence of physical geography topics demonstrates the influx of Vidal de la Blache's thinking about the unity of people and nature. Geography therefore adopts a descriptive character, such as the typology of landscapes in third and sixth years; and is explanatory about the major spatial processes and challenges of the present-day world in fourth, fifth and sixth years. This favours the student's constructive and abstract thinking by means of the concept of regional synthesis. Nevertheless, it is contradictorily complemented by a study by heart of geography (geographical places, terminology, etc.), which is designed as a follow on to the content learned at primary school. It mainly refers to relief, rivers, oceans and bioclimatic territories, both global and French.

In the instance of higher secondary education (*lycée*), there are two remarkable issues: geography and history are still taught as one single indivisible subject (even though the curriculum states that the content should be dealt alternatively, as in *collège*); and geography is taught in the three years (second, first, *terminale* for the 15, 16 and 17 year old) and in the three modalities of the general *Baccalaureat* (social, economic, literary and scientific). This means that most French students do geography for seven consecutive years. However, the big geographical topics in basic secondary education, like landscape, population, rural space, urban space, sustainable development, globalization, regional planning, French geography, etc. are dealt with again, but now with a higher degree of difficulty and a better specification of the objectives, skills and techniques. Thus, in general terms, the content of the 5th year is dealt with again in the 2nd year; the content of the 3rd year in the 1st year and 4th year content in the terminal year, paying special attention to globalization. Finally, at this level, the programme of study includes two areas of innovative content: i) a scientific focus, which includes two outstanding elements: the use of digital cartography and GIS; and ii) a spatial approach as an essential issue to understand the geopolitical and geo-economic keys of the present-day world, so that spatial citizenship and spatial thinking are clearly part of the French curriculum, although not explicitly expressed.

The Geography Curriculum in Finland

The Finnish curriculum in basic school education is quite schematic, in terms of objectives, content and assessment criteria, leaving enough space for curricular choices to schools and teachers. This issue has been proved by numerous studies highlighting the pedagogical autonomy of secondary schools, together with the high quality standards of teacher training as one of the successful factors of this educational system in international rankings. The approach allows teachers to be more involved in the constructivist approach when adapting content, curricular materials and didactic resources, learning and teaching methods, assessment tools, the social and educational contexts of the school, the characteristics of the class and the diversity among students.

In the case of geography, the curriculum for the last stage of basic education (Grades 7 to 9) is based on clarity and brevity. Yet, it establishes the assessment criteria for the end of the stage with no ambiguities. The blocks of content work by gradation of scale: the Earth, Europe and Finland, including aspects related to physical, human and regional geography in the three levels. The curriculum ends with a fourth block related to the environment and sustainable development. In total, there are fourteen sections of content referred to knowledge, since procedural and attitudinal content is included in the following section.

On the other hand, the section on assessment criteria is a bit longer than the one on content (18 in total) but they become a permanent reference for teachers in their learning and teaching practice. They are put together in five blocks of content: the first is about the acquisition of geographical skills (cartographical, graphical, statistical) and treatment of spatial information, and four more referred to the four blocks of content (Grades 7 to 9). The competencies are subjacent in the assessment criteria. Besides spatial competences (perception, description, comparison, explanation, analysis, etc.) there are others mentioned, such as social competence.

In higher secondary education the same phenomenon happens: there are general objectives, content (more developed here) and criteria, linked to the blocks of content. These comprise: a blue world (physical geography), a common world (cultural geography, population, settlements and urbanisation, economic geography, town and urban planning and sustainable development), a world of threats (natural, environmental, urban, demographic, and social). The fourth block is quite similar to the British curriculum as it suggests that case studies should be performed by means of a learning process based on a school research project, as a kind of regional synthesis, which should include the following features: collection and treatment of geographical information (literary, graphical, cartographical and statistical), proficiency of geographical working methods and presentation of conclusions using GIS and geo-media. This outline consists of the identification and relation of the geographical area to its national and international contexts, physical and human description, inner territorial structure and assessment of challenges and opportunities. For this work, the Finnish geography education system considers that is imperative to use geoinfomation and tools like PaikkaOppi, as shown in Chapter Six.

Finally, geography is considered important enough in Finland to be a national curriculum subject studied by all students throughout their secondary schooling. This illustrates the recognition of its importance for the educational and personal development of young adults in European society.

Geography Curriculum reform in Spain

At the time of writing this chapter, a new project to reform Spanish Educational Law and the National Curriculum, has been passed. Its implementation will entail the opportunity to revise, adapt and improve the school curriculum both in compulsory secondary education and in upper secondary education. Regarding geographical studies, two types of task should be considered: i) a critical analysis of different curricular options of geography in Spanish secondary education (following a diachronic method which will enable the evolution of the official curriculum to be checked), to change it and improve it; and ii) an innovative proposal to include into the curriculum the four challenges cited above, and mainly spatial thinking and spatial citizenship.

In the first instance, a wide bibliography already exists about the study of the curriculum of geography in Spain. The main evolutionary studies on the curriculum of geography have been carried out by Souto (2003, 2004, 2011) and González Gallego (2001, 2011), as well successive annual Conferences on Geographic Education held by the Group of Geographic Education of the Spanish Geographers Association.

The overall geography content in basic secondary education (ages 12-15) has not changed much, but rather their organisation or distribution from the first to third years. A somewhat classical outline of physical geography/human geography/regional geography has been followed. Nevertheless, in the second and third years more cross-curricular content is added to social sciences and its function of civic education: the current world, its features and problems, globalisation, spatial imbalances in development, political conflicts, cultural diversity, environmental challenges and sustainability, social inequalities, migration, etc. For this reason the curriculum in force is a bit more innovative and its approach allows the psychological passage of students from the concrete to the formal. Thus, students improve their spatial thinking from descriptions/locations to explanations/characterisations and eventually to interpretations of geographical phenomena and processes.

In turn, the geography curriculum in upper secondary education is only studied in the last year of school, after two years (ages 16 and 17) without learning geography. As a result, they suffer the same problems as in compulsory education, which can be verified after textbook research (De Miguel, 2013). The inclusions of common core contents (geographical working methods) as well as the incorporation of the block "Spain and Europe in the world" have not provided a necessary curriculum component. Apart from these issues, the content in upper secondary school is excessively orientated towards regional geography "anchored to a thematic academicism and a methodological conservatism" (learning by heart, lectures and minimal use of active methodologies), "expressed in some scholastic objectives and contents with little space for innovation" (Souto, 2011), which has produced the perception of geography in schools as being closer to a descriptive science rather than an explanatory science.

This is why, in any effort to review the curriculum, it is necessary that social and land issues, which are most current nowadays, should be incorporated in the teaching and learning of geography in schools. By including updated content that helps understand contemporary social phenomena or world conflicts, student interest in the subject will be raised. Changing is necessary to the present curriculum as it only suggests the use of geo-media through phrases like "new information and communication technologies". Like other countries previously analysed, the new Spanish curriculum must include expressions like geoinformation, GIS, spatial citizenship and geo-media. There are currently numerous innovative experiences in using geo-media to improve the learning and teaching of geography in Spain. Some of them were presented at the 2012 International Conference on Geographical Education (De Miguel et. al., 2012).

Conclusions

From the curricular analysis undertaken in this chapter, most of the countries researched have a similar framework in basic education and geography is present in all school years and cycles for students up to the age of 14-15. The main differences occur in upper secondary school, where only students in Spain do not get the opportunity to study geography. In most countries geography can be studied as an optional subject by all students, no matter what subjects they choose, whether humanities and social sciences, scientific or technological studies.

A distinction becomes necessary between i) the countries in the Anglo-Saxon and Nordic areas, where geography is an autonomous subject, detached from history, and ii) Mediterranean countries. The British, German and Finnish curricula list their own 'geographical' spatial competences like spatial thinking and spatial citizenship (related, somehow, to other kinds of competences, i.e. social, digital, etc.). On the other hand, the epistemological tradition of a regionalist geography, the teaching of geography combined with history, are factors which explain why France and Spain have not established a curriculum connected with specific spatial competences.

This distinction can be also applied to the very structure of the curriculum, excessively constrained to conceptual content, in the Spanish instance. This connection hinders the use of inquiry-based learning, case study approaches and geospatial activities on computers that are central aspects to the renewal of geographical education through the acquisition of spatial competencies.

In this way, in many of the curricula analysed above, the skills and methods relevant to geographical work are present, as well as the collection, treatment and expression of the geographical information in its four major aspects: literary, graphical, statistical, and, especially cartographical. These incorporate the use of new technologies in learning geography, for instance GIS and the possibilities that are opened up by the integration of the use of geoinformation and geo-media resources in secondary education. The next challenge will therefore be to increase the presence of geoinformation use in the national curricula, for example, through the final exam at the end of secondary education, and to include it into the "real curriculum" made by the teachers in high schools, autonomously to the textbook-curriculum developed by publishers.

The idea that geography education is a social science is reflected in all the curricula researched, even in Finland, where it is predominantly linked to biological studies. This supposes that geography favours the acquisition of social competences, and of course approaches like spatial citizenship, but specifically that it contributes to the students' maturity and overall studies so that they can put into effect their rights and duties in active life. For example, the exercise of democratic citizenship as well as the acquisition of a civic awareness, based in values such as human rights, and the joint responsibility in having a fair and equitable society. The research indicates that the British proposals condition the understanding of social phenomena and the acquisition of roles, values and attitudes to a research process, which shall contain personal perceptions of geographical phenomena, and which includes the assessment of spatial facts. On the other hand, the German approach goes beyond the raising of awareness, and, therefore includes competencies and abilities with the intention that students should evaluate and act coherently in environmental and social situations.

In the case of active methodologies, all the curricula reviewed are much more explicit than the Spanish one when it comes to drawing up inquiry-based approaches, problem-based learning, case studies, inductive methods, comparative analysis of different spaces in the same scale or spatial problems of different scale. The reform needs to establish places for more inquiry and learning activities of discovery, more prominence to fieldwork, a closer approach of the geography lessons as a geographical laboratory and as simulation of geographical research.

This was one of the main aspects to be learnt from this curricular review: replacing the rote study of geography for a reflective analytical study of the world: hence developments should focus on the importance of terms present in the assessment criteria of the British or Finnish curricula such as understanding, comparing, explaining and analysing rather than the terms identifying, defining, describing, found in the assessment criteria of the Spanish or French curriculum. That is to say: the geography curriculum should be regarded as a tool to induce cognitive processes rather than as a programme of academic content, as happens in the German curriculum in the instance of the areas of spatial competence and the skills related (Figure 2.1). In other words, the challenge is not teaching 'space' but rather teaching how to think about the space, and so, teaching with GIS rather than about GIS. It involves looking into the way students think of space as an abstract cognitive entity as this allows a better understanding of the learning processes related to the representation and appropriation of space.

Innovation in the geography curriculum in European countries will improve the quality and status of school geography and a better social regard for geographical education. Improving geographical education standards is needed, as it is one of the key subjects in scientific and social lifelong learning.

References

Bednarz, S. W. & Van der Schee, J. A. 2006. "Europe and the United States: The implementation of Geographical Information Systems in secondary education in two contexts". *Technology, Pedagogy and Education*, 15, 191-205.

Curic, Z., Vuk, R. & Jakovcic, M. 2007. "Geography curricula for compulsory education in 11 European countries – comparative analysis". *Methodical*, 15, pp. 467-493.

De Miguel, R. 2013. "Geoinformación e innovation en la enseñanza-aprendizaje de la Geografía: un reto pendiente en los libros de texto de secundaria", *Didáctica de las Ciencias Experimentales y Sociales* nº 27.

De Miguel, R., De Lázaro, M.L. & Marrón, M.J. (Eds.). 2012. *La educación geográfica digital*, Zaragoza, Grupo de Didáctica de la Geografía (A.G.E.) and Universidad de Zaragoza.

De Miguel. R. 2011. "Visores cartográficos y sistemas de información geográfica para la enseñanza y el aprendizaje de la Geografía en educación secundaria" in Delgado, J., de Lázaro, M.L. and Marrón, M.J. (Eds.) *Aportaciones de la Geografía en el aprendizaje a lo largo de la vida*. Málaga: Universidad de Málaga- Grupo de Didáctica de la Geografía (AGE), pp. 371-388.

Donert, K., ed. 2010. *Using Geoinformation in European Geography education*. Vol. IX. Rome: International Geographic Union-Home of Geography.

Donert, K., Parkinson, A. & Lindner-Fally, M. 2010. *Curriculum Opportunities for GeoInformation in Europe*, report from digital-earth.eu (SIG 4).

Donert, K. 2007. *Aspects of the State of Geography in European higher education*, Liverpool, Herodot network.

German Geographical Society. 2007 (2nd Edition at 2012), *Educational Standards in Geography for the Intermediate School Certificate*. 2007. Bonn: Deutsche Gesellschaft für Geography.

González Gallego, I. 2011. "Análisis crítico de las opciones curriculares en la educación secundaria obligatoria" in Prats, J., (Ed.), *Geografía e Historia. Complementos de formación disciplinar*. Barcelona: Graó, pp. 161-186.

—. 2001. "La formación inicial y permanente del profesorado de Geografía: una necesidad y un reto en el momento actual" in Marrón, M.J., (Ed.) *La formación geográfica de los ciudadanos en el cambio de milenio*. Madrid: AGE - Universidad Complutense de Madrid, pp. 673-701.

Gryl, I., Jekel, T. & Donert, K. 2010. "GI & Spatial Citizenship." in Jekel, T., Koller, A., Donert, K. & Vogler, R. (Ed.) *Learning with GI V*, ed., Berlin: Wichmann, pp. 2-11.

Houtsonen, L. 2006. "GIS in the school curriculum: pedagogical viewpoints" in (ed.) *Geographical Information Systems Applications for Schools*, Helsinki: University of Helsinki, pp. 23-29.

National Research Council (NRC). 2006. *Learning to think spatially. GIS as a Support System in the K-12 curriculum*. Washington, DC: National Academies Press.

Souto, X.M. 2011. "Fines y objetivos en la enseñanza de la Geografía: los condicionantes sociales y epistemológicos" in Prats, J., coord., *Didáctica de la Geografía e Historia*. Barcelona: Graó, pp. 115-129.

—. 2004. "La Geografía escolar en el periodo 1990-2003" in *Comité Español de la UGI, La Geografía española ante los retos de la sociedad actual. Aportación española al XXX Congreso de la UGI*. Glasgow, pp. 61-82.

Vogler, R., Henning, S., Jekel, T. & Donert, K. 2012. "Towards a concept of spatially enabled learning" in Jekel, T., Car, A. Strobl, J. and Griesebner, G. (Eds.) *GI Forum 2012: Revisualization, society and learning*. Berlin: Wichmann, pp. 272-282.

Whewell, C. P., Brooks, C., Butt, G. & Thurston, A. (Eds.). 2011. *Curriculum making in Geography: Edited conference proceedings of the International Geography Union Congress on Geography Education British Sub-committee 2011 Symposium*, held at Institute of Education, London, April 18th-20th 2011. London: Institute of Education, University of London and International Geographical Union Commission on Geographical Education.

CHAPTER THREE

THE NEED FOR A LEARNING LINE FOR SPATIAL THINKING USING GIS IN EDUCATION

LUC ZWARTJES

Spatial thinking & spatial literacy in Geography education

Today education is overwhelmed with all kinds of literacy: language literacy, mathematics literacy, computer literacy, technological literacy, science literacy, critical literacy, social literacy, relational literacy and many more. So what is so special about the need for spatial literacy?

Spatial literacy is defined as a set of abilities related to working and reasoning in a spatial world, like the ability to communicate in the form of a map, understand and recognise the world as viewed from above, recognise and interpret patterns, know that geography is more than just a list of places on the earth's surface, see the value of geography as a basis for organizing and discovering information, and comprehend such basic concepts as scale and spatial resolution (Goodchild, 2006).

The National Research Council formulates in their standard work 'Learning to think spatially' (Down et. al., 2006): a spatially literate person has following characteristics:

- He has the habit of mind of thinking spatially—he knows where, when, how and why to think spatially,
- He can practice spatial thinking in an informal way—he has deep and broad knowledge of spatial concepts (such as distance, direction, scale, and arrangement and representation (maps, 3D-models, and graphs…)
- He can adopt critical stance to spatial thinking and evaluate quality of spatial data, he can use spatial data to construct, articulate.

Spatial thinking is integral to everyday life. With the use of online mapping tools, GPS and car navigation the general public has become aware the possibilities made available by the use of spatial data. Additionally, it is the concept of space that makes spatial thinking a distinct form of thought. It is a basic and essential skill that can and should be learned, besides other skills like language, mathematics and science.

According to the National Research Council (Down et al. 2006) thinking spatially entails knowing about

- Spatial concepts—different ways of calculating distance, coordinate system and the nature of spaces in two and three dimensions. It includes also relative location, concepts of adjacency, intersections and regions.
- Spatial representation—the relationship among views: orthogonal versus perspective maps, the effect of projections, the principles of graphical design (semiology, Figure 3.2).
- Spatial reasoning—different ways of thinking about shortest distances, the ability to extrapolate and interpolate, estimate the slope of a hill from a map of contour lines...

For Michel & Hof (2013, Figure 1) it is exactly the links between these three features that gives spatial thinking its power of versatility and applicability.

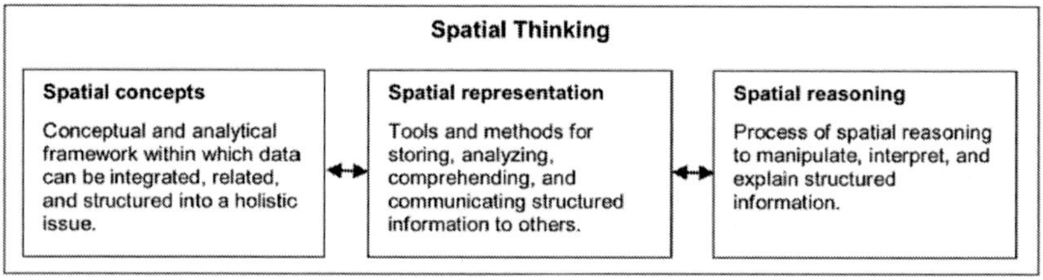

Figure 3.1. The concept of spatial thinking. (Michel & Hof, 2013)

Kerski (2008) summarises spatial thinking as the ability to study the characteristics and the interconnected processes of nature and human impact in time and at appropriate scales. In fact this is real geography: to be able to think critically about the earth, the activities of people and the interaction between the two. Thinking spatially is more than knowing where things are located, it's about asking geographic questions: why there, how originated and what if...?

Bednarz and Lee (2011) conclude, in their paper on the spatial thinking ability test (STAT), that spatial thinking is not a single ability but comprise a collection of different skills, whereby the following spatial thinking components emerge: map visualization and overlay, identification and classification of map symbols (point, line, area), use of Boolean operations, map navigation and recognition of spatial correlation.

Geographic skills provide necessary tools and techniques to think spatially. They enable us to observe patterns, recognise associations and consider spatial order. "Geographic representations ... are essential because they assist in visualizing spatial arrangements and patterns" (National Geography Standard, 2012).

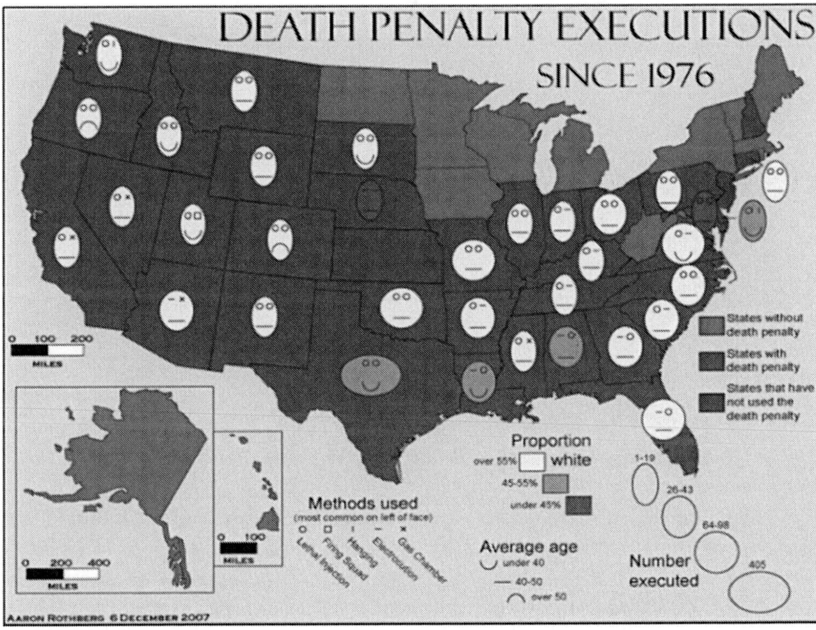

Figure 3.2. An incorrect use of semiology can give strange results. (http://wiki.ead.pucv.cl/index.php/Archivo:02_ejemplo_cartografia_penademuerte_chernoff.jpg)

In 2010 The U.S. Department of Labor developed a Geospatial Technology Competency Model (Figure 3.3.) "by researching and analysing publicly available resources, existing skill standards, competency-based curricula and certifications to provide an employer-driven framework of the skills needed for success in geospatial technology" (United States Department of Labor, 2010).

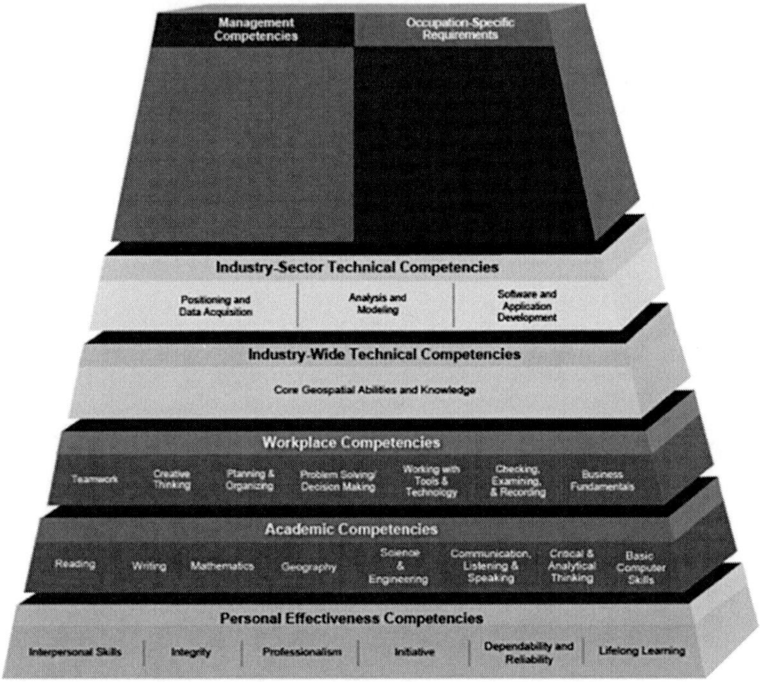

Figure 3.3. Geospatial Technology Competency Model. (http://www.careeronestop.org/competencymodel/)

Competency models offer job seekers, including students, an opportunity to learn what it takes to enter a particular field. For this competency model at the level of "academic competencies" geography is mentioned as "Understanding the science of place and space. Knowing how to ask and discover where things are located on the surface of the earth, why they are located where they are, how places differ from one another, and how people interact with the environment". (United States Department of Labor, 2010). If we examine more closely the specifications on level of skills and perspectives two concepts are conspicuously present: GIS and spatial thinking (Figure 3.4.).

Geography

Understanding the science of place and space. Knowing how to ask and discover where things are located on the surface of the earth, why they are located where they are, how places differ from one another, and how people interact with the environment.

- **Subject-specific Geographic Knowledge**
  - Human-Environment Interaction: Know and apply geographic information about relationships between nature and society (e.g., pollution from industrial development, economic effects of drought)
  - Regional Geography: Know and apply knowledge of the physical and human geography of a specific country or world region
  - Physical Geography: Know and apply geographic information about the processes that shape physical landscapes; weather, climate and atmospheric processes; ecosystems and ecological processes; and natural hazards
  - Cultural Geography: Know and apply geographic information about culture and cultural processes, including religion, language, ethnicity, diffusion, meaning of landscapes, cultural significance of place
- **Geographic Skills**
  - Geographic Information Systems (GIS): Use GIS to acquire, manage, display, and analyze spatial data in digital form
  - Cartography: Producing, creating, and designing paper or digital maps
  - Field Methods: Use interviews, questionnaires, observations, photography, maps, GPS, GIS, and other techniques to measure geographic information in the field
  - Spatial Statistics: Use quantitative methods to process spatial data for the purpose of making calculations, models, and inferences about space, spatial patterns, and spatial relationships
- **Geographic Perspectives**
  - Spatial Thinking: Identify, explain, and find meaning in spatial patterns and relationships, such as site conditions, how places are similar and different, the influence of a land feature on its neighbors, the nature of transitions between places, how places are linked at local, regional, and/or global scales
  - Global Perspective: Possess and apply knowledge of how people, places, and regions are linked by global networks and processes (e.g., globalization, international trade, immigration, Internet technology, global climate system)
  - Interdisciplinary Perspective: Draw on and synthesize the information, concepts, and methods of the natural and social sciences for geographic research and applications

Figure 3.4. Academic competencies of Geography inside the Geospatial Technology Competency Model.
(http://www.careeronestop.org/competencymodel/)

The use of GIS also plays an important role in acquiring geographic information literacy by combining geographic literacy (knowledge about geography) with information literacy (information search strategies, critical evaluation of sources) (Figure 3.5.). The outcome is the possession of concepts, abilities, and habits of mind (emotional dispositions) that allow an individual to understand and use geographic information properly and to participate more fully in the public debate about geography-related issues (Miller and Keller, 2005).

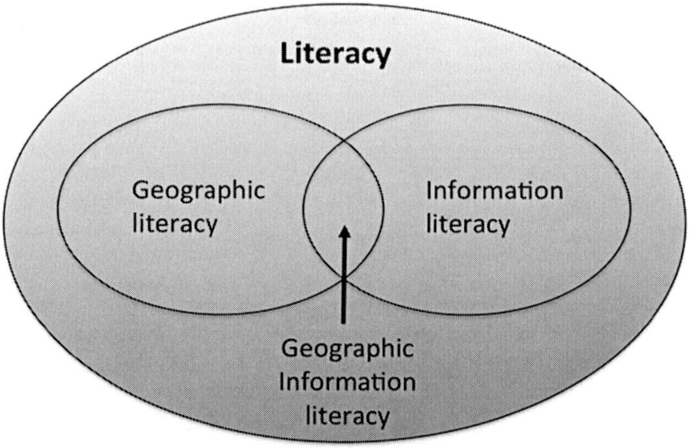

Figure 3.5. Contextual diagram for geographic information literacy. (Miller and Keller, 2005)

Linking spatial thinking & spatial literacy in Geography education

GIS (a Geographic Information System) is a set of computer technologies that allows the visualisation and manipulation of geographic data in an easy visual method. But GIS can also be termed "Geographic Information Science" (Goodschild, 1992), thus also involving a series of methods and ways of looking at the world (Milson, 2012), whereby GIS is used to obtain spatial thinking skills.

Freeman (1997) stated "changes in technology pervade the pedagogy and methodology of geography" so with the possibilities offered to use GIS nowadays (free software, available datasets, and computers with Internet access) we can no longer ignore the use of it in education. Koutsopoulos (2010) mentions two approaches for using GIS in education, he suggests:

- We can use the powers of GIS to teach geography, for they can help us to understand our world through both the natural and the man-made manifestations which are the essence of geography;
- In teaching with GIS a positive effect can be created on the development of spatial thinking and reasoning.

Thompson (1991) states that GIS is an "educational delivery system for improving the student's knowledge of the world in which she or he lives." GIS is able to answer all the questions that knowledge, understanding and application in Geography education require (Koutsopoulos, 2010). Thus "GIS can be defined as the study of the fundamental issues of geographic information, and is often motivated by the need to improve geographic information technologies" (Goodchild, 2011).

Because of its capabilities GIS is inherently an excellent vehicle in expressing the five themes of geography, as defined by the Joint Committee on Geographic Education (1984), namely: location, place, relationships with places, movement and region.

Koutsopoulos (2010) developed a conceptual framework for using GIS. As a basis, he uses the Geographic Education Standards Project (GESP, 1994), stating that geography is composed of three components: skills, subject matter and perspectives whereby all three are necessary if students are to be 'geographically informed' and thus should be examined (Figure 3.6.).

Geographic skills are a series of tools and techniques, including asking geographic questions, and acquiring and organising spatial information. The purpose is mainly focused on the level of knowing ("where is it?"), although some questions will lead to the process of understanding ("why is it there?") or even applying ("what if…?").

The *subject matter* is divided —according to GESP—into six "essential elements", most of which refer to the process of understanding.

A *geographic perspective* is a lens through which geographers look at the world. It involves the ways that knowledge and understanding can be used to solve geographic problems (process of applying). The specific aspect of geography—linking human and physical systems in a spatial lens—provides everything to solve spatial problems by active participation.

Geographic skills, subject matter and perspectives correspond to the processes of knowing, understanding and applying: by "learning the concepts and vocabulary of geography (knowing) students may begin to think about what they mean (understanding) and apply to real problems (applying)" (NAEP Geography Consensus Project, 2010).

Knowing is in spatial terms expressed by the questions, "What is it?" and "Where is it?" in GIS this means processing spatial data.

Understanding is expressed by questions such as: "Why is it there?", "What has changed?", "What is the pattern?", and "What is the interaction?" in GIS this is spatial analysis.

Applying is expressed by the question, "What if ...?" to solve spatial problems, in GIS this means planning.

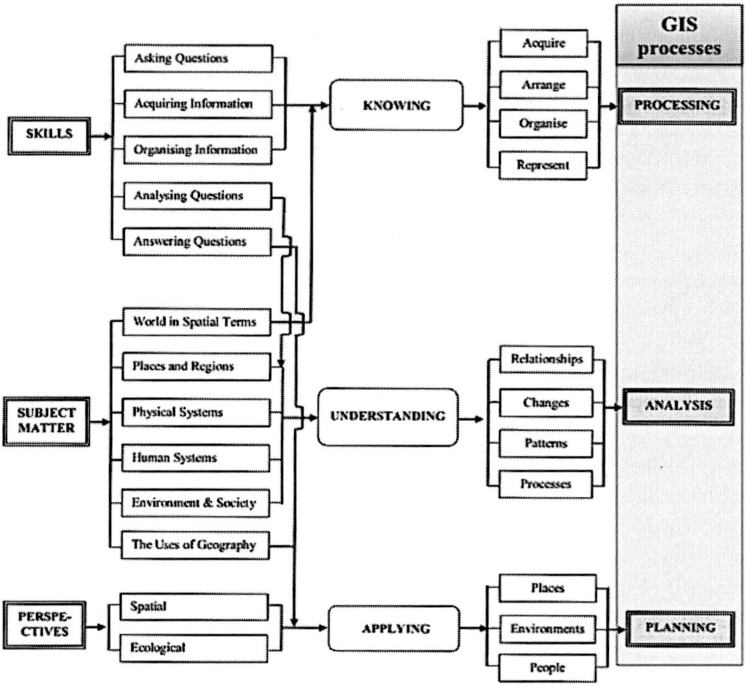

Figure 3.6. Linking the science of Geography to GIS – instructing with GIS. (Koutsopoulos, 2010 & own edit)

Koutsopoulos (2010) linked the three GIS processes with the questions and the five themes of geography—created by the Joint Committee on Geographic Education (1984): location, place, relationships with places, movement, and region (Figure 3.7.).

His framework very clearly shows the impact and importance of GIS in answering the questions at the level of the three processes. He proposes that "GIS can serve as an unique educational tool in which the manipulation, analysis and presentation of spatial data can support the teaching of geography" (Koutsopoulos, 2010).

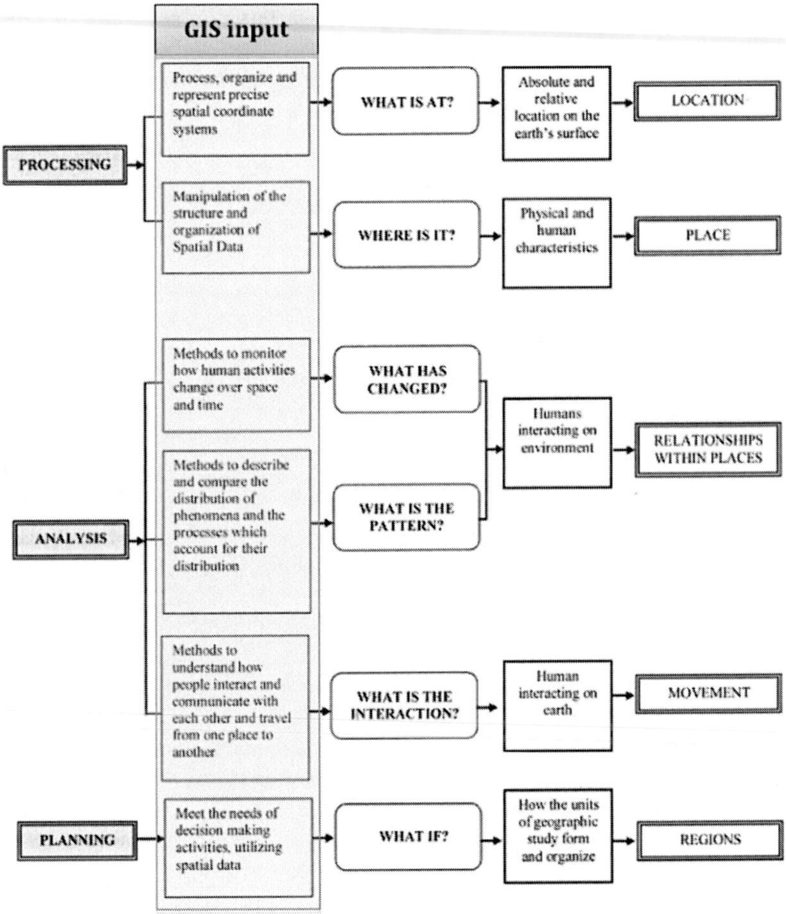

Figure 3.7. A conceptual framework in Instructing about GIS. (Koutsopoulos, 2010 & own edit)

More specific, typical spatial thinking skills are enhanced using GIS. By involving student activities using GIS "students not only learn by hearing and seeing, they also have the ability and opportunity to personally apply the knowledge using higher-order skills such as problem solving and synthesis" (Sanders, 2002). In order to foster such skills teachers and students may need to work in new ways such as through enquiry based methods and problem-based learning.

The approach developed by Koutsopoulos follows one of the four GIS schools described by Kemp (1992, quoted in Sui, 1995, Figure 3.8.): GIS as an enabling Technology for Science, arguing that GIS is not a goal in itself but a means to use spatial thinking skills.

Implications for GIS Instructional Approaches			
Relationship between Geography and GIS	*Instructional Focus*	*Course content*	*Format of delivery*
Geography as home discipline of GIS	education	application	combination of lab and lecture
GIS as a collection of marketable skills	training	technical	lab
GIS as a new tool for spatial science	training	application	lab
GIS as a new discipline	education	technical	lecture

Figure 3.8. Four schools of thought about the relationship between Geography & GIS (Kemp. et al, mentioned by Sui, 1995).

Two of the four schools describe the ideal vision for secondary education:

- The first schools state that geography is uniquely suited as the home discipline of GIS. It simply automates the tasks geographers have been doing for several thousands of years, and aims at a full integration of GIS into all aspects of geography curriculum.
- The third school envisions GIS as the tool to support scientific inquiry as an ultimate goal in a variety of disciplines, thus GIS is considered as an enabling tool for science.

Both of these places the emphasis of the course content on application—GIS as a tool, whereas the two other schools focus on the more technical aspects of GIS.

Integrating spatial thinking—using GIS- in education

The introduction of GIS in education has been argued with three complementary rationales that correspond to GIS's strengths:

1) The educative rationale: GI Science and GIS support the teaching and learning of geography;
2) The place-based rationale: GIS is the ideal tool to use to study geographical problems at a range of scales;

3) The workplace rationale: GIS is an essential tool for knowledge workers in the twenty-first century.

Van Leeuwen and Scholten (2009) see an added value of using GIS based on 5 senses:

- Sense of reality: using realistic data—e.g. of the own environment - makes abstract spatial theories become real
- Sense of urgency: by using realistic data and thematic items students get interested.
- Sense of experience of having influence: using GIS students get the opportunity to visualise a today's and tomorrow's landscape, influenced by (their) own decisions
- Sense of fun: people learn more easily when they are enjoying what they are doing and using GIS is fun when the tools are easy, interesting data is available and the case study is exiting.
- Sense of location: by using GIS in combination with GPS routes, tracking and tracing games or doing field work gives an extra dimension, location (x,y,z coordinates) becomes an exciting thing to explore.

However, these arguments have not appealed to large numbers of teachers. According Bednarz and Van der Schee (2006) the main reasons for this are:

- In teacher training (pre-service and in-service) GIS is not a core item.
- Non-geographers, tend to be teachers with limited pedagogical content knowledge, resulting in fewer teachers recognizing the potential opportunities GIS offers to teach geography content and skills, teach more and more geography.
- The curriculum doesn't include or impede adoption to include GIS.
- The availability of free data and easy-to-use software.
- The attitude of teachers. It seems difficult to persuade teachers to use new technologies, certainly if they are highly technically demanding and if teachers are not fully convinced of the effectiveness and added value that would result.

They made 3 recommendations,

1) Address the key internal issues related to GIS implementation: teacher training, availability of user-friendly software, and ICT equipment in schools

This was a matter of developing easier to use software with data access. As Goodchild (2011) concludes in his analysis of GIS software programs: "the GIS user interface remains complex, hard to learn and use, and lacking in any consistent conceptual or theoretical framework".

Nevertheless, a lot of progress has been made. There are many free GIS viewers (which eliminate the need to install software) and full, open source, GIS software programs available. Schools are nowadays well equipped with computers and a high speed (mobile) Internet. As a result of the INSPIRE directive more and more governments are offering datasets (for free) or provide open access to database servers. In different countries specific educational GIS-frameworks have been developed, like EduGIS in the Netherlands (Van der Schee, J. et al., 2006), the Pairform@nce Project in France (Genevois, 2011) or PaikkaOppi in Finland. Each of these learning environments offer users a simplified viewer—mostly inside a browser—with content that fits into the existing national curriculum.

The 2011 digital-earth.eu network survey on teacher training (Lindner-Fally and Zwartjes, 2012) concluded that only 45% of the participants have geoinformation/GIS included in teacher education/training in their countries, 55 per cent of teachers have to be provided with teacher education/training courses and information on available offers (Figure 3.9.).

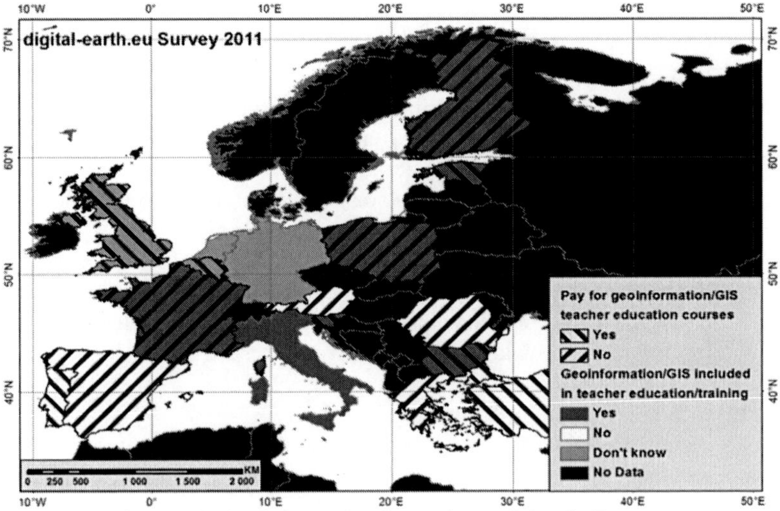

Figure 3.9. Geoinformation in teacher training in Europe. (digital-earth.eu survey 2011)

The impact of an ongoing form of teacher professional training is important. After attending teacher training in digital geo-media the teachers should be able to focus their work on the content rather than learning about the technologies. Bartocheck and Carlos (2013) showed that with greater frequency and longer training those who attended achieved higher levels of integration.

2) Use a community of learners approach.

A community of learners means bringing together, within a school or school region, all involved and crucial stakeholders in the educational process. Together they reflect and act upon best practices. Although this is a much praised and effective method, reality shows that certainly in secondary education this does not always work.

The digital-earth.eu network has accredited and launched "Centres of Excellence" in many countries. These organisations can help building up the community of geo-media learners and teachers, e.g. by collecting and disseminating good practice examples, organizing informal sessions with teachers.

3) Institutionalise GIS into curricula, making sure that it is aligned with significant general learning goals like graphicacy, critical thinking and citizenship skills.

This is also mentioned by The National Academy of Science (Downs et. al. 2006) who stated as one of the primordial recommendations the development of spatial thinking standards and curriculum material.

Favier (2013) describes five ways GIS can be integrated in secondary education (Figure 3.10.). Teaching and learning about GIS focuses more on the theoretical aspects of GIS (knowledge of GIS, structure of the technology), where the three other ways use the technology to develop and use spatial thinking skills.

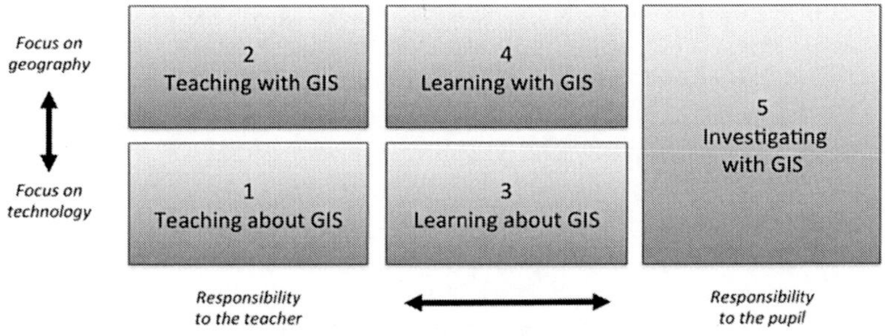

Figure 3.10. Five ways of integrating GIS in Geography education (Favier, 2013).

Research shows that most "successful" and easiest integration of GIS is done by "Investigating with GIS", where students are asked to do a real geographic enquiry. Liu and Zhu (2008) explain this by linking GIS to constructivism. Geographical enquiry draws on constructivism, emphasizing problem-solving and inquiry-based learning instead of instructional sequences for learning content skills. So GIS provides useful tools for constructing a computer-based, constructivist learning environment for geography education.

Without questioning the importance of this we must nevertheless try to generate a more continuous integration of GIS in education, using all five ways. The Irish pilot project "GIS into schools" is a good attempt to create and test curriculum materials for teaching GIS principles and practice (Tschirner and O'Brien, 2006). They indicate—just like Koutsopoulos (2010) and Favier (2013)—that to achieve an overall integration of GIS students first need to learn about GIS (theory and practice) and then apply this knowledge to learn with GIS. The Irish example used several geography curriculum-based topics and was thus not really integrated over the different years of study.

Creating GIS learning outcomes through education

The digital-earth.eu project examines the use of geographic media in schools and teacher education. Geo-media is the visualisation of information from different media sources and is concerned with digital content and its processing based on place, position and location. Many geographic media are widely used for navigation and routeing purposes. The digital-earth.eu network seeks to provide broad access to resources, promoting innovation and best practice in the implementation of geo-media as a digital learning environment for school learning and teaching. The goal is to raise the profile of learning with digital earth tools and resources. The network encourages the sharing of innovative practices and rewards organizations and individuals displaying "excellence".

Special Interest Groups (SIGs) work on following topics

1. Resources, technologies and geoinformation;
2. Learning and teaching with geo-media and geoinformation;
3. Teacher Education and Training in geo-media;
4. Curriculum aspects and geo-media.

Developing spatial literacy assumes the availability and use of digital earth tools, which allow students to interact with geoinformation, to answer questions and critically reflect using a geographic approach. They can also clearly communicate the results to a broader audience.

Special Interest Group 3 of the digital-earth.eu network addressed the importance of introducing GIS (use of geo-media) for three competences (Figure 3.11.) (Woloszynska et al. 2013).

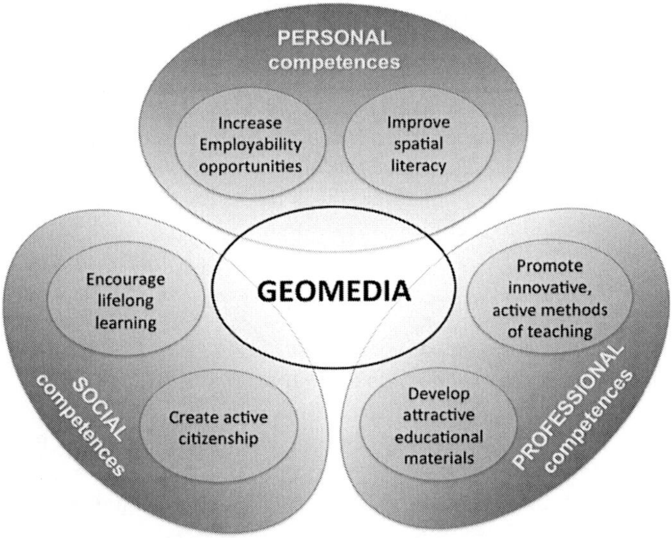

Figure 3.11. Why geo-media in teacher training (Woloszynska et al. 2013).

Personal competences

Developing spatial literacy assumes interaction with geoinformation. A geographic approach is necessary to answer questions critically and constructively. Teachers must therefore understand basic geographic concepts and be able to support students' learning needs. Employability is enhanced by geo-media skills.

Social competences

Education for active citizenship equips people with the content knowledge, skills and understanding to play an effective role in society. They become interested in controversial issues and engaged in discussion, debate and decision-making. Therefore, education for spatial citizenship plays an important role for the learning process.

To enable teachers to bridge the technological gap between students and themselves, they need to use geo-media in the classroom to allow learners to explore real world issues and encourage lifelong learning strategies.

Professional competences

Geo-media brings the real world into the classroom. Constructive and active learning practices like problem solving, project-based learning, fieldwork strategies and enquiry approaches are favoured and will help them to face future challenges.

Therefore, teachers must understand basic geographic concepts and be able to support students' learning needs. Taking in account the different levels of age and education, teachers must be enabled to apply different methods and tools in the respective learning environments.

To help promote this a benchmark has been developed[1], indicating the competencies needed to achieve spatial literacy.

Competencies:

- Spatial thinking:
    - To know concepts of spatial thinking;
    - Be able to use tools of spatial representation;
    - To apply processes of reasoning (Where is it? Why is it there? What if it was somewhere else? Making informed decisions and defend personal points of view);
- Pedagogic and didactical skills for the use of digital earth tools in school;
- Ability to use spatial skills in real world problem-solving context;
- Understanding complex and changing interrelationships;
- Awareness and understanding for the digital earth concept;
- Ability to use digital earth tools (also technological skills);
- Lifelong learning competencies: ability to find training opportunities, time management, planning competency, and communication competencies;
- Being able to identify and evaluate resources;
- Social learning:

o Being able to work with others – teamwork;

o Use professional social networks (virtual and face-to-face).

In order to prepare teachers to effectively implement digital earth in their practice, teacher training and teacher education therefore needs to appropriately prepare teachers for different levels of education.

---

[1] This benchmark statement has been produced as a result of the digital-earth.eu COMENIUS network SIG 3 (Teacher education and teacher training) meeting in Bruges, Belgium in October 2011. http://www.digital-earth.eu

Primary school teachers need to be able to enable students (year 1-6) to

- Open digital maps and virtual globes on a computer;
- Indicate the different parts of digital maps/virtual globes (navigation bar, menu, scale, map window, Figure 3.12.);
- Interpret symbols on digital maps;
- Work with digital maps and 3D representations of the world:
    - o Find significant locations (their home, school or town) on a virtual globe;
    - o Pan, zoom, orientate;
    - o Make measurements;
    - o Use the layers to focus on specific features;
    - o Update maps;
- Be aware of generalization levels applied in different zoom levels (e.g. road density);
- Access information efficiently and effectively, evaluate information critically and competently (see maps as manipulated representations created by people/organizations with a certain purpose, e.g. classification methods, colour schemes, map contents);
- Use digital maps and virtual globes for a variety of different purposes.

Figure 3.12. Primary school pupils should be able to work with digital globes and simple GIS-software.

Secondary school (year 7-12)

In addition to the learning outcomes of primary school, secondary school teachers need to enable their students to:

- Know the digital earth concept and its tools;

- Understand the basic purpose and application of digital earth to real world problems;
- Be able to gather and evaluate information;
- Use advanced digital earth tools for learning (starting with Web-GIS, GIS viewers to GIS software);
- Manipulate maps:
    - o Display information on maps;
    - o Create own maps;
    - o Communicate cartographic information;
- Understand the construction of digital maps as a representation of the real world:
    - o The power of maps (reliability of data, classification and colour schemes);
    - o Topology: points, lines, and polygons;
    - o Layers;
    - o Database;
- Know about the professional use of GIS and other digital earth tools;
- Gather information from data resources or through fieldwork activities (use GPS devices, mobile applications);
- Use digital earth tools for investigation/research:
    - o Interpret content
    - o Identify and ask significant questions that clarify various points of view and lead to sustainable solutions;
    - o Frame, analyse and synthesize information in order to solve problems and answer questions.

Creating a learning line in GIS

Taking into account the age of students and level of complexity of activities, the best option for curriculum development is to create a learning line, covering at least the six years of school study, from age 12 to 18. It would be even better if GIS activities started in primary school education.

A learning line is defined here as the educational term that refers to the construction of knowledge and skills throughout the whole curriculum. It reflects an increasing level of complexity, ranging from easy (more basic skills and knowledge) to more difficult.

As an example of this the Flemish Geography National Curriculum (LEERPLANCOMMISSIE AARDRIJKSKUNDE (2010) defines these learning lines for Secondary Schools:

5 learning lines:	Fieldwork	Working with images	Working with maps	Working with statistical material	Creation of knowledge
Level 1	**Perception** – knowledge of facts				
Level 2	**Analysis** – selection of relevant geographic information				
Level 3	**Structure** – look for complex connections and relationships				
Level 4	**Apply** – thinking problem solving				

Some existing examples:
Learning line maps:

- Level 1: Recognise and name the elements of the legend on the map. Distract the scale;
- Level 2: Retrieve from the map those geographic elements that are relevant within a research context;
- Level 3: Classify and relate elements on the geographic map;
- Level 4: Interpret a map;

Learning line images:

- Level 1: Describe the image;
- Level 2: Retrieve from the image those geographic elements that are relevant within a research context;
- Level 3: Examine the correlation between the different elements by using various techniques (map studies, surveys, statistics...);
- Level 4: Make up a synthesis of the image.

When applying the learning line concept to the learning outcomes (described in the previous section) we get this result:

Level 1: Perception—being able to work with digital maps and virtual globes:

- Open digital maps and virtual globes on a computer;
- Indicate the different parts of digital maps/virtual globes (navigation bar, menu, scale, map window);
- Interpret symbols on digital maps;
- Understand the construction of digital maps as a representation of the real world (topology, layers, database).

Level 2: Analysis—selection of the relevant geographic information:

- Work with digital maps and virtual globes: find locations, pan, zoom, orientate, make measurements;
- Access information efficiently and effectively, evaluate information critically and competently;
- Be able to gather and evaluate information from data resources or through fieldwork activities;
- Interpret content.

Level 3: Structure—look for complex connections and relationships:

- Use digital maps and virtual globes for a variety of different purposes;
- Identify and ask significant questions that clarify various points of view and lead to sustainable solutions;
- Manipulate maps by creating own maps;
- Communicate cartographic information.

Level 4: Apply—thinking problem solving:

- Be aware of generalization levels applied in different zoom levels (e.g. road density);
- Understand the basic purpose and application of digital earth to real world problems;
- Use advanced digital earth tools for learning (starting with Web-GIS, GIS viewers to GIS software);
- Frame, analyse and synthesise information in order to solve problems and answer questions.

For introduction in the different grades of schools the level would depend on the age. Level 1 should be reached in primary education; level 2 can already be reached in primary—depending of the class group—but must be reached in lower secondary education. Level 3 can be reached in lower secondary – again depending of the class group, but must be reached together with level 4 in upper secondary education.

	Level 1 Perception	Level 2 Analysis	Level 3 Structure	Level 4 Apply
Primary (6-10 y)				
Lower Secondary (11-14 y)				
Upper Secondary (15-18 y)				

Another method which could be adopted is the cataloguing of the competencies into competence areas used as the basis for the development of a learning line (Woloszynska et. al. 2013), whereby teachers should be able to choose suitable tools to use based on the abilities and needs of their students, their own capabilities and their curriculum.

Competence Areas	**Primary** **6 – 10 y**	**Lower Secondary** **11 – 14 y** *(In addition to 6-10 y)*	**Upper Secondary** **15 – 18 y** *(In addition to 11-14 y)*
understanding / analysing digital geomedia	reading, orientating, combining, interpreting, measuring, comparing, querying		geo processing network analysis spatial analysis
producing and communicating digital geomedia	collaborative activities, mapping, visualising, sharing, discussing update geo-media, maps, infographics, charts, presentations collect and represent information add information to maps and other geo-media thematic mapping *... at different levels of scale and complexity over the years*		
critical use / awareness of digital geomedia in everyday life	awareness of generalization, different zoom levels, perspectives, intentions, manipulated representations, volunteered geographical information (vgi) reflect on content and representation, information rights and ethics identification of digital media in everyday life geomedia as part of decision making		
geographical technology: hardware & tools	GPS, digital maps, virtual globes web mapping	3D representations of the world (DEM) satellite images open geodata online, desktop and mobile GIS	

Conclusions

Spatial thinking should be a compulsory feature of school education like linguistic and mathematical thinking. Because of its capabilities, GIS is inherently an excellent vehicle to deliver essential spatial thinking skills. The framework developed by Koutsopoulos (2010) shows very clearly the impact and importance of GIS in answering the questions on the level of knowledge, understanding and application. In this respect, GIS can serve as a unique educational tool in which the manipulation, analysis and presentation of spatial data can support the teaching of geography.

The introduction of GIS in secondary education is not easy. Many of the reasons why earlier attempts to incorporate GIS in geographical education have not succeeded have been overcome by recent developments. But most teachers will only be persuaded once GIS implementation is included in the curriculum. The benchmark created by the digital-earth.eu project is a first step as it offers a framework on which a curriculum learning line could be established. From this classroom content can be constructed, depending on the age of the pupil and real curriculum reform might result.

References

Bartoschek, T., Carlos, V., 2013, What happens when teacher training in digital geo-media is over? Case studies analysing levels of pedagogical integration, *GI_Forum 2013: Creating the GISociety* (Eds. Jekel, T., Car, A., Strobl, J., Griesebner, G.), Wichmann Verlag Berlin, 437-446.

Bednarz, R. and Lee, J., 2011, The components of spatial thinking: empirical evidence, *Procedia Social and Behavorial Sciences* 21, 103-107

Bednarz, S. and Baker, T., 2003, Lessons learned from reviewing research in GIS education, *Journal of Geography (National Council for Geographic Education)*, 102 (6), 231-233.

Bednarz, S. and Van Der Schee, J., 2006, Europe and the United States: the implementation of geographic information systems in secondary education in two contexts, *Journal of Technology, Pedagogy and Education*, Vol. 15, No. 2, 191-205.

Chun, B.A., Hong, I., 2007, Integrating GIS with Geographic and Environmental Education into K-12: an Interdisciplinary Curriculum Development Entitled Studying the Environment of Eighteen mile Creek, *Journal of the Korean Geographical Society*, Vol.42, No.2, 295-313.

Dibiase, D. et. al., 2006, *Geographic Information Science and Technology Body of Knowledge*, Association of American Geographers, 162 p.

Downs, R.M. (chair), 2006, *Learning to think spatially: GIS as a Support System in the K-12 Curriculum*, National Research Council, National Academy Press, 313 p.

Favier, T.M., 2011, *Geographic Information Systems in inquiry-based secondary Geography education*, Vrije Universiteit Amsterdam, 287 p.

—. 2013, *Geo-informationtechnologie in het voortgezet aardrijkskundeonderwijs: Een brochure voor docent,* Vrije Universiteit Amsterdam, 80 p.

Freeman, D., 1997, Using Information Technology and New Technologies in Geography, *Teaching and Learning Geography*, London: Rutledge, 202-217.

Joint Committee On Geographic Education, 1984, *Guidelines for Geographic Education: Elementary and Secondary Schools*. Washington DC, 28 p.

Generous, S., 2011, Teacher training for using Geo Information in secondary education – Experiences from the Pairform@nce Project, *GI_Forum 2011: Learning with GI 2011* (Eds. Jekel, T., Koller, A., Donert, K., Vogler, R.) Wichmann Verlag Berlin, 104-107.

Geography Education Standards Project, 1994, *Geography for Life – National Geography Standards*, National Geographic Society, Washington D.C., 272 p.

—. 2012, *Geography for Life – National Geography Standards, Second edition*, National Geographic Society, Washington D.C., 272 p.

Goodchild, M.F., 1992, Geographic information science. *International Journal of Geographical Information Systems*, Vol. 6, 3-45

—. 2006, *The Fourth R? Rethinking GIS Education*, ArcNews. Fall 2006

—. 2011, Spatial thinking and the GIS User Interface. *Procedia Social and Behavioral Sciences* 21 (2011), 3-9

Janelle, D.G., Goodchild, M.F., 2009, Location across Disciplines: Reflections on the CSISS Experience, *Geospatial Technology and the role of location in science* (Ed. Scholten, H., van de Veldt, R., van Manen, N.), Springer, Dordrecht, 15-29

Johnson, A.B., 2001, PedagogicalApproaches to Teaching GIS: "Hook them or Sink Them", *The 4th AGILE Conference on GIScience*, Brno

Kerski, J., 2003, The implementation and Effectiveness of Geographic Information Systems Technology and Methods in Secondary Education, *Journal of Geography (National Council for Geographic Education)* 102, 128-137

—. 2008, *Developing Spatial Thinking Skills in Education and Society*, ESRI ArcWatch January 2008, http://www.esri.com/news/arcwatch/0108/spatial-thinking.html

Kotsopoulos, K., 2010, Teaching Geography – Instructing with GIS and about GIS, *Using GeoInformation in European Geography education*, 1-19.

Leerplancommissie Aardrijkskunde, 2010, *Aardrijkskunde Tweed Grad ASO*, VVKSO, Brussel. http://ond.vvkso-ict.com/vvksomainnieuw/document.asp?DocID=2431

Lindner-Fally, M., Zwartjes, L., 2012, Learning and teaching with Digital Earth – Teacher training and education in Europe, *GI_Forum 2012:*

Geovizualisation, Society and Learning (Eds. Jekel, T., Car, A., Strobl, J., Griesebner, G.), Wichmann Verlag Berlin, 272-282

Liu, S., Xian, Z., 2008, Designing a structured and Interactive learning environment based on GIS for secondary Geography education, *Journal of Geography (National Council for Geographic Education)* 107, 12-19

Marais, H.J.W., 2008, *The Challenges of GIS Education and Training*: (GIS Use by Municipal Urban and Regional Planning), http://www.a-a-r-s.org/acrs/proceedings2008.php

Michel, E., Hof, A., 2013, Promoting Spatial Thinking and Learning with Mobile Field Trips and ego-Riddles, *GI_Forum 2013: Creating the GISociety* (Eds. Jekel, T., Car, A., Strobl, J., Griesebner, G.), Wichmann Verlag Berlin, 378-387

Miller, J., Keller, P.C., 2005, Suggested Geographic Information Literacy for K-12, *International Research in Geographical and Environmental Education*, Vol 14, No.4, 243-260.

Milson, J.M., et al., 2012, *International perspectives on teaching and learning with GIS in secondary schools*, Springer Science + Business Media B.V., New York – Heidelberg, 353 p.

Naep Geography Consensus Project, 2010, *Geography Framework for the 2010 National Assessment of Educational Progress*, U.S. Department of Education, Washington D.C., 72 p., http://www.nagb.org/content/nagb/assets/documents/publications/frameworks/gframework2010.pdf

Sanders, L.R., teal. 2002, Electronic mapping in Education, *Journal of Research on Technology in Education*, Vol.34, No.2, 91-1009

Sui, D., 1995, A pedagogic framework to link GIS to the intellectual core of Geography, *Journal of Geography (National Council for Geographic Education)*, Vol. 94, N° 6, 578-591.

Thompson, D., 1991, GIS – A view from the other (dark) side: the perspective of an instructor, Introductory Geography Courses at University Level, *Cartographical*, Vol. 28, N° 3, 55-64

Tshirner, S., O'Brien, M., 2006, GIS into schools: developing a secondary GIS curriculum http://citeseerx.ist.psu.edu/viewdoc/download?doi=10.1.1.100.4843&rep=rep1&type=pdf

United States Department Of Labor, 2010, *US Department of Labor announces release of Geospatial Technology Competency Model*, United States Department of Labor, http://www.dol.gov/opa/media/press/eta/eta20100950.htm.

Van Der Schee, J., Korevaar, W, Scholten, H.J., 2006, Aardrijkskundeonderwijs draagt met GIS bij aan ruimtelijk inzicht, *Geo-informatica* juni-juli 2006, 33-36.

Van Der Schee, J., Scholten, H.J., 2009, Geographical Information Systems and Geography Teaching, *Geospatial Technology and the role of location in science* (Ed. Scholten, H., van de Veldt, R., van Manen, N.), Springer, Dordrecht, 287-301

Van Leuven, W.S., Scholten, H.J., 2009, Spatial literacy: the ABC of the (X,Y,Z). The five senses of GIS in education, *Global Spatial Data Infrastructure Association*, http://www.gsdi.org/gsdi11/papers/pdf/186.pdf

Woloszynska, E., Carlos, V., Carvoeiras, L., Cosac, M., De Lazaro, M., De Miguel, R., De Mira, H., Lambrinos, N., Lang, B., Linder-Fally, M., Martinha, C., Schmeinck, D, Silva, D., Zwartjes, L., 2013, *Teaching with Digital-Earth. Guidance for teacher trainers*, Digital-Earth.eu network, www.digital-earth.eu.

Zwartjes, L., 2012, Creating a learning line in education, *GIS-education: Where are the boundaries? - 8th European GIS Education Seminar proceedings* (Hubeau, M., de Bakker, M., Toppan, F., Reinhardt, W., Steenberghen, T., Van Orshoven, J. Eds.) EUGISES, Leuven, http://ees.kuleuven.be/eugises12/eugises12-seminar-proceedings.pdf.

CHAPTER FOUR

DIGITAL EARTH AND GEOGRAPHY TEACHER TRAINING FOR THE 21ST CENTURY: TEACHER COMPETENCIES FOR INQUIRY-BASED GEOGRAPHY TEACHING

TIM FAVIER AND JOOP VAN DER SCHEE

Introduction

Planet earth is more and more becoming a digital earth. Hundreds of millions of people worldwide regularly access and interact with geoinformation via powerful technologies, such as web-GIS, online atlases and virtual globes, geo-information based smart phone apps. Many digital media use geo-referenced information in our physical and social environment, and these also belong to the category of geo-media. Digital-earth.eu is a network of organisations, which aims to stimulate the use of geo-media in school education (Jekel, Koller, Donert & Vogler, 2011).

Geographic Information Systems (GIS) is a very important category of geo-media application and is one of the fastest growing technologies. GIS can be seen as a kind of software that offers access to large sets of geoinformation (GI). GI is information with a spatial component, and is used by tools to visualise, manipulate, query, analyse, and present complex situations in a fast and flexible way. For this reason, GIS forms an ideal instrument for studying geographic problems. In the past decade, more and more geography teachers, teacher trainers, and scientists in the field of geography education have become interested in the possibilities of using GIS in secondary geography education. GIS offers especially many opportunities to set up inquiry projects in which students investigate geographic problems in authentic contexts (NRC, 2006). In such a way, GIS enables teachers to make their lessons more interesting and challenging for students. However, research conducted by Kerski in 2003 showed that the introduction and spread of GIS in secondary education was only proceeding slowly. Eight years later, the use of GIS in inquiry-based education was still very limited in most countries (Milson, Demirci & Kerski, 2011). One of the frequently cited barriers to the integration of GIS in teaching is teachers' limited knowledge about how to use GIS in teaching and learning (Bednarz & Van der Schee, 2006). However, there is little understanding of the nature of this problem. Little is known about what knowledge teachers should have to design and coach good inquiry projects with GIS, and what knowledge in practice forms the major bottleneck.

The central question this chapter addresses is: What knowledge should teachers have to design and coach viable and effective geographic inquiry projects with GIS? This research is a follow-up of an earlier Computers & Education paper (Favier & Van der Schee, 2012) that focused on the characteristics of an optimal design for GIS-supported geographic inquiry projects. Both papers discuss the outcomes of a PhD research (Favier, 2011) that aimed to provide insight in how GIS-supported inquiry-based geography education can be realised.

Theoretical considerations

Those who want to investigate the knowledge basis of teachers for GIS-supported inquiry-based geography education are inevitably confronted with two questions. The first question is: "What is the core of 'geographic literacy'?" Besides this, a second question arises: "What is understood by inquiry-based education?" Geographic literacy refers to the knowledge and skills that are needed to understand and study problems in the world around us (Van der Schee, 2007). Not just knowledge, but knowledge about problems in which the location on earth matters. It is, for example, knowledge about natural disasters, water management, migration, urban planning, globalisation and climate change. Geographic literacy refers to skills such as selecting, reading, analyzing, interpreting and constructing maps, and skills in asking and answering geographic questions. Next to geographic knowledge and geographic skills, motivation also plays an important role: the willingness to understand the world around us; and the willingness to conduct geographic inquiry. So, together with geographic literacy, "geographic drive" (Favier, 2011) is also needed.

The second question is, what exactly is meant by geographic inquiry? This term refers to the activities that are conducted to gain more understanding of problems in the world around us. Empirical geographic inquiry follows, just like other empirical inquiry, the well-known cycle of formulating questions, collecting information, processing information, answering questions, and formulating new questions.

Research approach

To investigate how GIS-supported inquiry-based geography education can be realised, a research approach was followed that falls under the heading of design-based research, also called Educational Design Research (EDR) or design research (McKenney, Nieveen & Van den Akker, 2006). The research team consisted of nine teachers from six schools in different parts of the Netherlands, and the first author of this paper, hereafter called "the researcher". None of the teachers had experience in teaching with GIS. Together with the researcher, the teachers developed a geographic inquiry project for 4th and 5th grade HAVO (senior general education) and VWO (pre-university education) classes via five progressive cycles of designing, testing, and evaluating (Fig.1). The test rounds took place between 2008 and 2010. Most teachers conducted the project two or three times at their school. In total, 375 students participated in the tests. Most students were 15, 16 or 17 years old. The earlier paper (Favier & Van der Schee, 2012) describes the research approach in more detail.

In the course of the design-based research, the design of the inquiry project improved, and the teachers became more competent in conducting the project with their students. Gradually, the research team gained insight into the characteristics of a viable and effective design (Favier & Van der Schee, 2012), and insight into the knowledge that teachers should have to design and coach viable and effective geographic inquiry projects with GIS, and the knowledge that forms a mayor bottleneck to designing and coaching such projects.

In order to investigate what knowledge teachers need, and what knowledge forms a major bottleneck, the following data was collected during the design-based research: (1) videotapes of whole-class discussions at the start and end of the inquiry project; (2) videotapes of one-to-one (teacher-student) discussions during the inquiry project; (3) videotapes of meetings of the research team in which the teachers and researcher evaluated the tests and explored how to improve the design of the project; (4) surveys conducted among teachers; and (5) interviews conducted among teachers. The surveys and interviews focused on teachers' opinion about the knowledge they developed in the course of the design-based research, and the bottlenecks they experienced when designing, testing, and evaluating the inquiry project. The entire process was documented in a context-rich narrative description of the design activities, test activities, and evaluation activities (Favier, 2011). The next section describes the geographic inquiry project with GIS, and summarises the main conclusions of the design-based research with regard to the teacher knowledge basis.

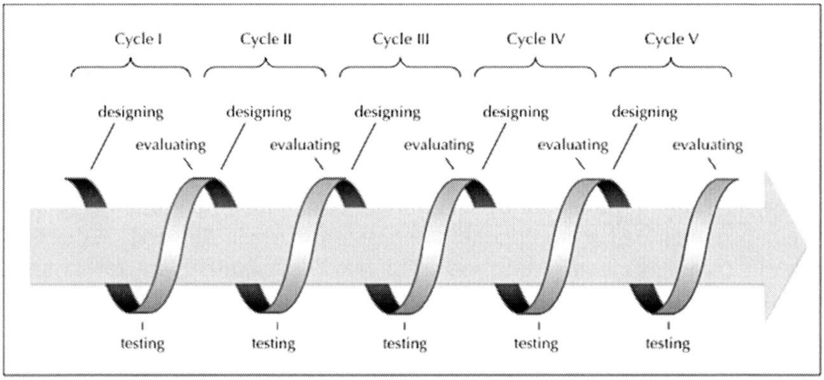

Figure 4.1. Set up of the design-based research (Favier, 2011).

Description of the GIS-supported geographic inquiry project

The inquiry project that was developed in the design-based research was named "Services & Customers". The aim of the project is to map the market areas of services such as cinemas, bakeries, swimming pools or fashion stores, and to determine which factors can explain the differences in the size of those market areas. Every service has its own market area. It is the area in which the customers of the service live. The inquiry project consists of 11 phases (Figure 4.2.), and takes about 8 one-hour lessons. It covers a full cycle of the geographic inquiry process. One of the most important phases is the inquiry-planning phase. In this phase, students choose 3 or 4 services and formulate hypotheses about the size of the market areas of those services and the factors that can explain those differences. They also have to think about which data they need to test their hypotheses, and construct surveys. Then they go to the services and interview 20 customers at each service. They ask, among other things, for the postal code of the customers, which can be used to map the market area. As they also have to explain differences in the size of market areas, they also have to ask additional survey questions. They can ask, for example, which mode of transport the customers used to get to the service, or why the customers chose to visit that service rather than another service. Back at school, students enter the data in Excel and use the Excel spreadsheet to construct digital GIS maps. Figure 3 shows an example of maps constructed in test round III by two students who focused their project on gyms in Gorinchem, a medium-sized town in the central part of the Netherlands. At the end of the inquiry project, students present the outcomes of their investigations to the other students and the teacher. A whole-class discussion follows in which the teacher and students try to summarise what was learned.

Phase		Activities
1	Introduction	The teacher explains the learning goals, set-up, and planning of the inquiry project.
2	GIS training	Students follow a GIS training in which the learn about the characteristics of GIS and geodata, and learn to use the relevant tools.
3	Preparation	Students make assignments in which they learn how to construct good surveys.
4	Inquiry planning	Students select three or four services, formule inquiry questions and hypotheses, and construct surveys.
5	Data collection	Students interview 20 customers at every service.
6	Data entering	Student enter the data in Excel.
7	Data visualisation	Students construct GIS maps on the basis of their Excel file.
8	Constructing presentations	Students construct presentations.
9	Presenting	Students present the outcomes of their inquiry project to their classmates and teacher.
10	Whole-class discussion	The teacher and students summarize what was learned from the inquiry project
11	Reporting	Students write a report about their inquiry project.

Figure 4.2. Setup of the geographic inquiry project with GIS (Favier, 2011).

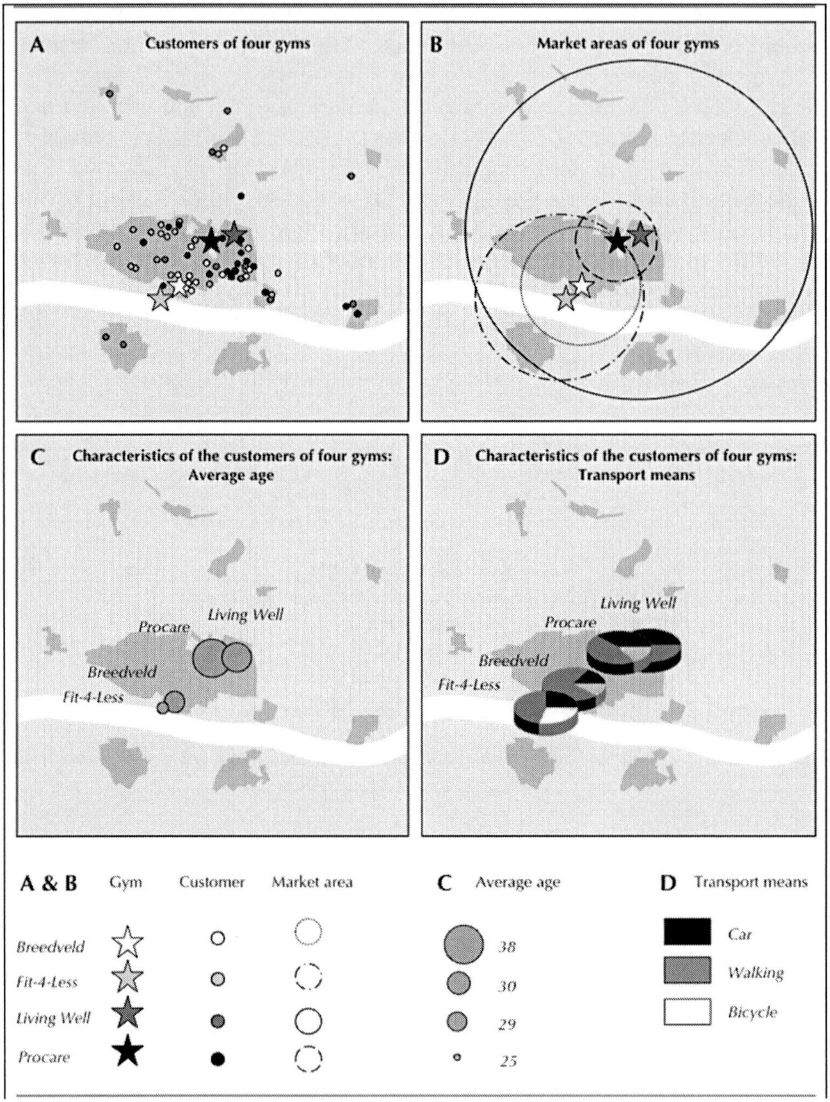

Figure 4.3. Simplified maps of two students who mapped the market areas of four gyms in Gorinchem, and investigated the factors that influence the size of those market areas (Favier, 2011).

Results

As the teachers had little or no GIS knowledge, they first followed a one-day GIS training module in which they learned about the structure of GIS and tried out different GIS tools. During the training, one of the teachers put forward the idea of mapping the distribution of customers of services on the basis of post code data, and mapping the market areas of services with the buffer tool. This was the basis for the first version of the GIS project, and the research team subsequently designed handouts for entering data in Excel and visualizing data in GIS. In the first test rounds of the design-based research, however, many students got stuck when they were working with GIS. The problems were mainly caused by omissions and errors in the handouts, by errors in students' Excel files, and by inaccurate reading of the handouts. Teachers were often not able to attribute the cause of the problems, and when they were able to diagnose problems correctly, they often did not know how to solve them. In order to keep the project going, the researcher offered help to the teachers and showed them the cause and possible solutions of the problems. In the course of the design-based research, the teachers were more and more able to help students themselves. In addition to this, after several cycles of trial-and-error, the teachers also learned how to design the tasks in the handouts in such a way that many problems were avoided. This means that teachers gradually developed the necessary GIS knowledge and GIS-didactic knowledge. As a result of the improvements in the design and progression in teachers' knowledge, the viability of the GIS project increased.

At the end of cycle III, when most technical problems were under control, the focus shifted to another problem: the teachers realised they missed a domain-specific framework that could be used to structure students' geographic thinking and raise it to a higher level. Below is a fragment of a presentation of the two students from Gorinchem who focused their project on gyms to illustrate this issue. During the presentation, the students argued that:

> *"When we started the project, we thought that young people visit gyms more frequently than old people. And it turned out to be exactly like that. Look at this map (Figure 4.3. C). For old people, the distance also plays a role of course, because the distance determines how far you want to travel. Old people don't want to travel far. For them, the quality is more important. You can see that at Procare. Procare is a nice gym with good coaching. They also have a sauna and physiotherapy. And classes like Callenetics, which aim at old people. And look: it has the highest average age of all gyms. So it is because of the quality that people choose Procare, although it is more expensive."*

This shows that the students discovered some important things, but that they did not discuss the geographic content in a structured way. First of all, their reasoning is a bit messy. For example, what did they mean when they said, “For old people, the distance also plays a role of course, because the distance determines how far you want to travel?” It is not clear which relationship they were talking about. Besides this, students focused on the relationship between the quality of a gym and the age of the customers, instead of focusing on the relationship between the quality of a gym and the size of market areas. Also, students overlooked an important relationship: they did not notice that Procare, the ‘posh’ gym, is situated in a neighbourhood with a very high average income. So it was very logical that most customers lived in the direct vicinity of the gym. The density of people with a high income was probably the main explanatory factor for the relatively small market area of Procare. Because the students did not analyze the geographic content in a structured way, they were hardly able to answer their research question. This shows that it is necessary that teachers coach students in structuring the geographic content, so that they develop geographic knowledge and skills in studying geographic problems more effectively. However, the teacher asked the students only one question after their presentation. And that is not typical for this teacher and this student pair. In the first couple of test rounds, the teachers hardly engaged in discussions with their students, while the presentations provided many opportunities to discuss the geographic content of the project and to link it with geographic theories from textbooks about spatial behaviour of consumers, the distribution of services, etc. Also, teachers hardly stimulated students to reflect on their inquiry approach, while the quality of students’ inquiry questions, hypotheses and surveys was generally low.

In the group discussion after test round III, it turned out that most teachers found it difficult to judge the content of the presentations. The main problem was, however, that the teachers did not have a sound domain-specific framework for use in educational settings in their mind to analyze and interpret the information. Teachers should have a sort of theory in their minds in which the relevant relationships and interactions are expressed in both verbal and visual form. Figure 4 shows the part of the theory that is needed to coach the two students from Gorinchem in structuring the geographic content. Teachers need to have such a theory in their minds in order to: (1) identify and interpret patterns and relationships in students’ data; (2) recognise and attribute inaccuracies in students’ reasoning; and (3) formulate appropriate interventions. So, most teachers missed the necessary geographic-didactic knowledge to raise the effectiveness of the GIS project. In order to fill this knowledge deficit, the teachers and researcher subsequently developed a framework (as is shown in Figure 4) and a list of potentially effective survey questions that are connected to this theory. During the group meeting, it turned out that most teachers had limited background knowledge of the relevant geographic theories about spatial behaviour of consumers. More important, it turned out that most teachers were not used to analyze geographic content and geographic inquiry methods in a structured way and to transform knowledge about this content and methods in order to make it suitable for use in educational settings. So, in other words, they missed the necessary geographic-didactic knowledge.

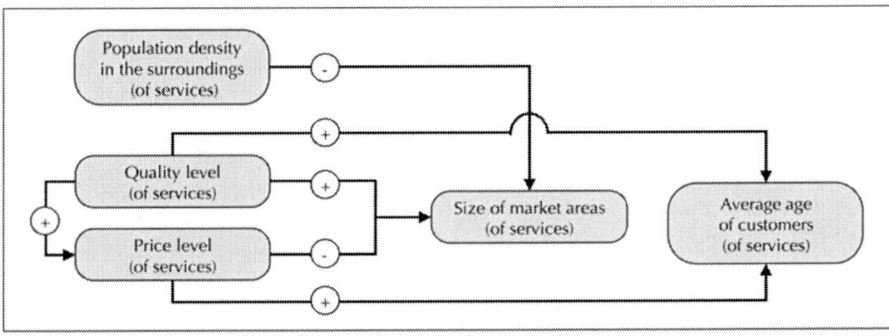

Figure 4.4. Part of the theory about the factors that influence the size of market areas of services (Favier, 2011).

Although the teachers found it difficult to construct a sound domain-specific framework for use in educational settings, they were very enthusiastic about het end product. Each of the nine teachers decided to use the framework. In the following two test rounds, the teachers paid much more attention on the geographic content and on appropriate geographic inquiry approaches. As they now had more geographic-didactic knowledge, they engaged more frequently in one-on-one (teacher-student) discussions. Also, most teachers organised an extensive whole-class discussion at the end of the project. However, some teachers still found it difficult to coach students in structuring the geographic content via dialogical teaching and to stimulate students to reflect on their inquiry strategy.

Inquiry-based education with GIS is a complex kind of education that requires a systematic approach. In the final two cycles of the design-based research, the research team included a couple of preparatory assignments in the set-up of the GIS project, such as an assignment in which students learn to formulate relevant and correct survey questions. The research team also included some assignments at the end of the project, such as an assignment in which students had to express the relationships in the form of rules and generalizations. Developing such assignments was a complex task. It required a couple of cycles of designing, testing and evaluating. Gradually, the teachers learned how students learn in relationship to such tasks, and used this knowledge to improve the tasks.

It is clear that the teachers who participated in the design-based research were not used to delving so deep in trying to improve their teaching and spending so much time and effort in transforming geographic knowledge. During the final research team meeting, the teachers stated that they very much appreciated the process of collaboratively designing, testing, and evaluating a GIS project. A survey conducted among the teachers after the final cycle of the design-based research confirmed this observation. The question "How useful was engaging in the design-based research for your professional development as a teacher?" resulted in an average score of 4.5 on a 1 to 5 Likert scale.

Conclusions

Geographic inquiry projects with GIS can have a large impact on the development of students' geographic literacy. However, such projects require a very systematic approach. In order to make sure that GIS-supported geographic inquiry projects are viable and effective, teachers should offer a considerable amount of guidance when students formulate inquiry plans, work with GIS, reason about the geographic content, and reflect on the inquiry strategy. This is very difficult and designing and coaching projects with GIS requires many competencies. The design-based research shows that teachers need to have sufficient GIS knowledge and GIS-didactic knowledge to secure the viability, and sufficient geographic knowledge and geographic-didactic knowledge to raise the effectiveness of geographic inquiry projects with GIS. Regarding geographic-didactic knowledge, it is especially important that teachers are able to transfer their geographic subject knowledge and geographic methodological knowledge to "domain-specific frameworks for use in educational settings", which is more a matter of restructuring knowledge rather than simplifying knowledge. This kind of geographic-didactic knowledge is not only needed for teachers who want to integrate GIS in their teaching, but also for everyone who wants to raise students' geographic thinking to a higher level. At the start of the design-based research described above, teachers' knowledge was insufficient to secure the viability and to maximise the effectiveness. However, the teachers were able to develop the required knowledge. By designing, testing, and evaluating a geographic inquiry project with GIS, together with other teachers and a researcher, they gradually learned how to design and coach a project in such a way that it is feasible for students as well as teachers and effectively contributes to the development of students' geographic literacy.

Discussion

It is clear that teaching geography with GIS is very complex. Teachers should therefore be trained via pre-service and in-service training. Such training programs should not only focus on the development of GIS-knowledge and GIS-didactic knowledge, but also on the development of geographic knowledge and geographic-didactic knowledge. However, learning to teach geography with GIS is also a matter of learning by experience. Teachers should therefore start with conducting simple GIS-supported geographic inquiry projects. Support from an expert who offers ideas for the set-up of the project and helps in diagnosing and solving problems during the GIS lessons is very useful for the development of teachers' geographic-didactic knowledge. In the USA, ESRI's geo-mentor programme aims to offer such in-class support. Although the help of GIS experts is very much appreciated by teachers, the main focus of the geo-mentor program is on the organizational, practical, and technical issues. In this paper, we have seen that teachers could also use some help with the didactical issues. It is therefore advisable that experienced and inexperienced teachers work together in a "community of learners" (Beishuizen, 2004) in which they explore how they can improve the quality of geography education with GIS. Some of the teachers who participated in the design-based research now have such an expert role within their school. They made their colleagues enthusiastic about teaching with GIS, and now support them in designing and conducting different kinds of GIS-supported geographic inquiry projects. Teachers are gatekeepers of educational changes and educational innovation. Together they can help students to discover planet earth using all kind of digital tools to prepare for today's a tomorrows citizenship.

References

Bednarz, S. W. & Van der Schee, J. A. (2006). Europe and the United States: The implementation of Geographical Information Systems in secondary education in two contexts. *Technology, Pedagogy and Education*, 15, 191-205.

Beishuizen, J. (2004). Computer-supported inquiry learning: Effects of training and practice. *Computers & Education*, 42, 389-402.

Favier, T.T. (2011). *Geographic Information Systems in inquiry-based secondary Geography education. Theory and practice*. Amsterdam: Vrije Universiteit (PhD Dissertation).

Favier, T.T. & Van der Schee, J.A. (2012). Exploring the characteristics of an optimal design for inquiry-based Geography education with Geographic Information Systems. *Computers & Education* 58, pp. 666-677.

Jekel, Th., Koller, A., Donert, K. & Vogler, R. (2011) Implementing Digital Earth in Education. In Th. Jekel, A. Koller, K. Donert & R. Vogler (Eds.) *Learning with GI 2011*. Berlin: Wichmann.

Kerski, J. J. (2003). The Implementation and Effectiveness of Geographic Information Systems Technology and Methods in Secondary Education. *Journal of Geography*, 102, 128-137.

McKenney, S., Niemen, N. & Ackers, J. van den (2006). Design research from a curriculum perspective. In Akker, J. van den, Gravemeijer, K., McKenney, S. & Nieveen, N. (Eds.) *Educational Design Research*, pp. 67-90. Oxon, UK: Routledge.

Milson, A.J., Demirci, A. & Kerski, J.J. (2011). *International perspectives on teaching and learning with GIS in secondary schools*. New York, NY: Springer.

NRC (2006). *Learning to think spatially: GIS as a support system in the K-12 curriculum*. Washington D.C.: National Academic Press.

Van der Schee, J. (2007). *Gosse leerlingen. Geografische Informatie Systemen, geografisch besef en aardrijkskundeonderwijs*. Amsterdam: Vrije Universiteit (orate).

CHAPTER FIVE

LEARNING AND TEACHING WITH GEOSPATIAL TECHNOLOGIES IN SPAIN

ISAAC BUZO, MARIA LUISA DE LÁZARO AND MARÍA DEL CARMEN MÍNGUEZ

Introduction

Increased digitisation of spatial information has raised the relative importance of geography in the curriculum. This should be taken advantage of by teachers and trainers in order to improve the spatial competences of students, and through which lifelong learning approaches can be encouraged. This chapter begins with an evaluation of the importance of training teachers in spatial and digital earth competences, after that it examines one particular learning experience based on the development of a Spanish web site for teachers and concludes with a rationale for the importance of the integrated use of geospatial technologies in education.

Geography in secondary schools in Spain

Since 1990, when the reform of secondary education in Spain was approved by the Law Regulating the General Educational System (LOGSE), Spanish secondary education has consisted of a compulsory period of four academic years (ESO: students aged from 12 to 16) and a post-compulsory period that can be two academic years of pre-university studies (*Bachillerato*: students aged from 16 to 18) or vocational education (Vocational Training). Since then legislative reforms have changed only specific aspects without affecting the general structure of the education system.

In compulsory secondary school, geography content is included in a global area named social sciences, geography and history. It also deals with other subjects such as history, history of art and economy, distributed as follows:

- 1st Academic year: physical geography with prehistory and ancient history.
- 2nd Academic year: medieval and modern history with human geography and population.
- 3rd Academic Year: purely geography content, both physical and human geography of the world with special emphasis on economic aspects.
- 4th Academic Year: contemporary history.

At the *Bachillerato* level (secondary non-compulsory schooling), geography is an independent subject in the second academic year. This is an optional subject to be taken by students whose profile is directed towards university studies in Law and Social Sciences (geography, economics, history, etc). It is orientated towards the study of the geography of Spain. Pupils who have studied geography in their 2nd year *Bachillerato* may select it as one of the subjects for the university entrance exams. During the four years of compulsory education, social sciences, geography and history take up 3 hours per week while in 2nd *Bachillerato* they are taught for 4 hours a week.

The importance of training teachers in spatial and digital earth competences

Teacher-training is essential for the transfer of innovation and improvement of geography teaching. With the introduction and development of new technologies this training should deal with competences in the use of GeoInformation (GI) and Geographical Information Technologies (GIT). In order to accomplish this, it would be very useful if standards for teacher training could be agreed so that they can be used to develop courses for GI and GIT in teaching geography. In some universities GIT is already being taught as a module for ICT, but it does not have a specific geographical or spatial perspective in line with UNESCO standards (2008). In geography, the official Spanish Master‘s in Secondary Education Teaching, Vocational Training and Language Teaching would provide such an opportunity.

Spatial data and Digital Earth resources for teachers

Web 2.0 provides teachers, trainers and students with the potential to find information about, access to and sharing of many spatial and digital earth tools and technologies. For instance, there are many very useful resources for teaching and learning geography available from different Spatial Data Infrastructure (SDI) portals, Virtual Globes (VG) and Web-based Geographical Information Systems (GIS). They can all prove very helpful to geography teachers, but institutional support is also necessary because of the large volume of new resources being made available in the current period of rapid change of geo-media development.

a) The Spatial Data Infrastructure (SDI) of the INSPIRE Directive of the EU, together with the standards of the Open Geospatial Consortium (OGC) has considerably increased the amount of available geospatial information on the Web. So far, this mainly relates to institutional geoinformation gathered by public bodies with links to available geoinformation in any state, local and regional governments according to their legal frameworks. It also provides detailed information on other databases. In Spain, many useful links have been collected and shared (González & Lázaro, 2011). However, it is important for teachers to learn how to use them for their own classroom and research purposes, for example when teaching and learning geography using Virtual Globes and GIS as interfaces.

b) Virtual globes (VG) provide a means to teach and learn geographical concepts and improve spatial competences (Jekel, T., Pree, J. & Kraxberger, V., 2007; Lázaro, González & Lozano, 2008; Schultz, R. B., Kerski, J. J., & Patterson, T. 2008). They have become very popular because of the visual nature of the product and the easy of use of the interface provided. This makes them relatively easy to introduce in teacher training courses and in school geography lessons.

c) Geographical Information Systems (GIS) can be used to ensure that students acquire spatial skills (Donert, 2005; Koutsopoulos, 2008). However, in a country like Spain, there are still not enough opportunities in the official school curriculum framework to learn and teach with GIS in secondary schools (Campo et al., 2011). This is despite the fact that European benchmarks for GIS competences exist, for example that created by the HERODOT network (http://www.herodot.net/Geography-benchmark.html), which should be reviewed and localised to fit a Spanish context.

d) Technological barriers have been minimised by the introduction of social cartography tools as an alternative to GIS software, for instance by using Map Maker or My Maps in Google, or Google Map Creator, a tool developed under the GeoVUE project (https://www.bartlett.ucl.ac.uk/casa/research/past-projects/geovue).

All of these resources made new methodologies possible in class situations, for example the development of collaborative mapping with tools such as Wikimapia, OpenStreetMap, Google Map Maker, Flickr Maps, GeoCommons, Eye on Earth or ArcGIS Online. Citizens are able to give feedback by gathering information for scientific projects. This type of collaboration has been implemented to generate large volumes of data as volunteered geographic information (VGI). The VGI approach is sometimes termed asserted geographic information (Goodchild, 2007) as it places the participant at the heart of a Citizen Science process. The GLOBE programme, initiated in 1995, is a good example. Social cartography, cartography based on experience, can also be used. The process of community participation in the research, data collecting experiences and the knowledge from systematic data management (physical evidence) can lead students to transform their involvement into action (Kerski, 2011). Web-based geospatial tools allow different types of interactive or hot maps to be created and shared, based on areas of conflict (Buzo, 2010).

Geography teaching and learning on a teacher's website: Experience in Extremadura (Spain)

Computerization of the classroom

Educational administration in Spain is shared between the state government, which establishes the basic law for education (curriculum), and regional governments, who manage schools. So every Spanish autonomous region has developed its own way to access learning and teaching resources online. In some cases, schools were explicitly designed with several specific computer classrooms which students can use whenever necessary. In other places, such as the Autonomous Community of Extremadura (Colmenero, 2011), it was decided to computerise all classrooms in all secondary schools in the region, so that a personal computer was available for every two students that worked with free software (a version of Linux called LinEx). Currently the Ministry for Education of the state government, in collaboration with regional governments, is implementing a new computer classroom system called "Escuela 2.0" whereby every secondary school student will receive a laptop computer to use both in the classroom and at home.

Despite the economic investment in recent years teachers are still not very receptive to the daily uses of ICT, given the amount of inconvenience that the use of computers still presents (not enough training, computer failure, resource availability on the network, etc.). Many teachers have recognised that online resources provide fundamental tools for 21st century education. A student who finishes secondary education should have acquired some specific digital skills. On the other hand it is a highly motivating tool for students and facilitates the teaching-learning process through ability to access information and perform complex tasks like simulations.

Web site used as a daily tool in Geography classes

Over the years a large website has been developed to support the work of teachers of geography <http://contenidos.educarex.es/sama/2010/csociales_geografia_historia/index.html>. The result has been the development of both demonstrations and interactive activities for teachers and pupils to work with every day in class. The content has been organised according to social sciences subjects, geography and history. So far, activities for social science in 1st, 2nd and 3rd year of compulsory secondary education (ESO) have been created and for geography in the 2nd year of the *Bachillerato*. The remaining subjects will be developed in subsequent years.

The objective is to get all students to pick up some digital skills by using the site. To do this, a variety of tools have been used in the preparation of activities. Some of them expose the content of a topic more or less interactively (e.g. Flash presentations). However, others are designed so that students learn the function of a tool while at the same time learning the content of the subject (e.g. drawing different graphs using a spreadsheet). Activities also vary according to the academic levels of students. In the first years of study, with younger students, activities are more interactive and varied. In the upper grades, where the pressure from the university entrance examination to finish the syllabus is greater, the activities involve more presentations, with a greater use of graphics and images. The different types of activities can be summarised as:

a) Flash-based (Buzo, 2010). These are interactive activities made by taking advantage of Adobe Flash software as a tool that allows multimedia files (sound, video and images in different formats) to be integrated, through which it is possible to interact, creating movement and providing textual information (controlling when and where the text and images will appear). The activities made using Flash are:

- Interactive maps: this type of representation is very useful for learning geographical locations and can be combined with game-based learning (hit locations, putting a geographical element in place...) to further motivate the students in the resolution of the activity—http://contenidos.educarex.es/sama/2010/csociales_geografia_historia/primeroeso/mapas/mapa_politico_europa.html;
- Lineal slideshows: used to present information slowly and in a specific order—http://contenidos.educarex.es/sama/2010/csociales_geografia_historia/primeroeso/tema2/origen_relieve.html;
- Slideshows with many buttons: in which the order of the information contained in the presentation is not important, thus allowing the student the opportunity to explore and choose the information they want to find—http://contenidos.educarex.es/sama/2010/csociales_geografia_historia/segundoeso/tema4/ciudad.html;
- Sensitive Maps: by passing the mouse over a particular location, information appears related to the place (name, definition, etc.). This type is useful in pointing out different parts on the screen—http://contenidos.educarex.es/sama/2010/csociales_geografia_historia/primeroeso/tema2/formas_relieve_costa.html;
- Chart analysis: applies the same technique used in sensitive maps to the analysis of graphics so that by passing the cursor over the graphical information further information is displayed—http://contenidos.educarex.es/sama/2010/csociales_geografia_historia/segundoeso/tema14/poblacion.html;
- Animated models representing reality: simple animations that can help explain processes that are associated with the idea of change or movement and may have some abstract component which may otherwise be difficult for the student to understand. This type of animation can be exported and converted to video format—http://contenidos.educarex.es/sama/2010/csociales_geografia_historia/primeroeso/tema3/lluvia.html.

b) Activities based on Web 2.0 tools such as presentations, maps, images, time lines, etc. These tools provide the opportunity to easily share work between and with students and other teachers. Moreover these activities are embedded in a simple way by using html code on a website or blog. Among the activities planned for this kind of web 2.0 tools, the following are noted:

- A series of more than 60 presentations posted on Slideshare on the geography of Spain and integrated onto the web of Social Science Resources in each of the subjects—
http://contenidos.educarex.es/sama/2010/csociales_geografia_historia/geografia/presentaciones/presentacion_factores_elementos_clima.html;
- Simple and explanatory videos of historical and geographical phenomena are uploaded to its own channel on YouTube and integrated in the respective subjects—
http://contenidos.educarex.es/sama/2010/csociales_geografia_historia/geografia/videos/video_coriolis.html;
- A series of pictures posted on Picasa, Flickr and Slide that illustrate exercises and examples—
http://contenidos.educarex.es/sama/2010/csociales_geografia_historia/fotografias/foto_mes4.html;
- Maps made with Google Maps for geo-referencing different images, events, etc.—
http://contenidos.educarex.es/sama/2010/csociales_geografia_historia/fotografias/foto_mes4.html;
- Timelines made with Dimity—
http://contenidos.educarex.es/sama/2010/csociales_geografia_historia/segundoeso/tema11/cronologia_america.html.

c) Activities developed with HTML technology. These are useful for embedding websites on other websites, or to produce sensitive maps. In these activities students can observe the contents of different websites at the same time on the screen, relating the contents of both. One example is the case of weather map exercises—
<http://contenidos.educarex.es/sama/2010/csociales_geografia_historia/primeroeso/tema3/mapa_tiempo.html>, which combines a sensitive map with an embedded frame. The student clicks on various elements of the map and descriptions appear inside the frame.

d) Activities that promote the active and advanced uses of office programs such as word processing, spreadsheet, presentation, drawing, so that students can use all these tools at the end of their schooling—
http://contenidos.educarex.es/sama/2010/csociales_geografia_historia/segundoeso/tema4/crecimiento_poblacion.html.

e) Social Networks. For communication with students and geography teachers. One group for teachers of geography on Facebook is at
http://www.facebook.com/group.php?gid=62146897459, a Facebook page to announce updates to the Resources for Social Sciences web pages

http://www.facebook.com/pages/Recursos-de-Ciencias-Sociales-Geograf%C3%ADa-e-Historia/294435090386 and a Twitter channel for communicating news on the web.

Conclusion

Geography teachers need to know how to make the best use of the many sources of geoinformation and geo-media available to them. In order to improve their cartographic skills they need be able to access quality spatial data, and then apply this through basic uses of virtual globes and Web-based Geographical Information Systems. With these tools, they will be able to introduce collaborative mapping and communication using Web 2.0 technologies, increasing active participation and motivation. Students must learn to recognise the real nature of place, its spatial perspectives and symbolism through map making. These activities will in turn provide them with digital and other professional skills.

References

Buzo, I. (2010), Posibilidades didácticas del Flash para la enseñanza de la Geografía. In Marrón, M.J. & Lázaro, M.L. (eds.), *Geografía, Educación y Formación del Profesorado en el Marco del Espacio Europeo de Educación Superior*. Madrid: Grupo de Didáctica de la Geografía (A.G.E.) and Universidad Complutense de Madrid, pp. 147-159.

—. (2011a), La cotidianeidad en el uso de las TIC en las Ciencias Sociales. In: Hernández, J., Pennesi, M., Sobrino, D. & Vázquez, A. (eds.) 2011. *Experiencias Educativas en las Aulas del Siglo XXI. Innovación educativa*. Barcelona: Ariel, pp 347-350.

—. (2011b), Recursos de Ciencias Sociales, Geografía e Historia. [online] Available in: <http://contenidos.educarex.es/sama/2010/csociales_geografia_historia/index.html>

Campo, A., Romera, C.; Capdevila, J.; Nieto, J.A. and Lázaro, M.L. (2012), Spain: Institutional Initiatives for Improving Geography Teaching with GIS, in *The World at their Fingertips: International Perspectives on Teaching and Learning with GIS in Secondary Schools*. Milson, A. J.; Kerski, J.J. & Demirci, A. Springer.

Colmenero, P., (2011), Introducción de las TIC en las aulas de secundaria de Extremadura. In: J. Hernández, M. Pennesi, D. Sobrino, A.

Vázquez, (Eds.) (2011). *Experiencias Educativas en las Aulas del Siglo XXI. Innovación educativa*. Barcelona: Ariel, pp 282-285.

Donert, K. ed. (2005), *Higher Education GIS in Geography: a European perspective*. Liverpool Hope University.

Goodchild, M. F. (2007), Citizens as sensors: the world of volunteered Geography. In *GeoJournal* 69 (4), pp.: 211-221. http://www.springerlink.com/content/h013jk125081j628/ (07/2012)

González, M.J. & Lázaro, M.L. (2011), La geoinformación y su importancia para las tecnologías de la información geográfica. *Ar@cne. Revista electrónica de recursos en Internet sobre Geografía y Ciencias Sociales*. Barcelona: Universidad de Barcelona, nº 148, 1 de junio de 2011. <http://www.ub.es/geocrit/aracne/aracne-148.htm>. (07/2012)

Jekel, T., Pree, J. and Kraxberger, V. (2007). Kollaborative Lernumgebungen mit digitalen Globen. Eine explorative Evaluation. In: Jekel, T., Koller, A. and Strobl, J., *Lernen mit Geoinformation II*. – Heidelberg: Wichmann, 116-126.

Koutsopoulos, K. (2008), The Case of Geoinformatics in Europeanization. In: *Journal of Geography in Higher Education* 32 (1):7-14.

Lázaro, M.L. and González, M.J. (2005), La utilidad de los Sistemas de Información geográfica para la enseñanza de la Geografía. In: *Didáctica Geográfica* 7, pp.105-122. <http://www.didacticageografica.es/> (11/2012)

Lázaro, M.L. and González, M.J. (2007), Spain on the web: A GIS way on teaching. In Herodot/Eurogeo. *Teaching in and about Europe. Geography in European higher education*. Toruń: Herodot/Eurogeo. vol. 4, pp. 36-43.

Lázaro, M.L., González, M.J. and Ruiz, M. E. (2007), Excursiones virtuales y migraciones. In: AGE. *Las competencias geográficas para la educación ciudadana*. Valencia: AGE- Grupo de Didáctica, p. 371-386.

Lázaro, M.L. and González, M.J. (2008), Learning about immigration in Spain through geoinformation on the Internet. In Donert, K.(Ed.). *Future Prospects in Geography*. Liverpool: Ed. Herodot/Liverpool Hope University Press, pp. 439-446.

Lázaro, M.L, González, M.J. and Lozano, M.J. (2008), Google Earth and ArcGIS Explorer in Geographical Education. In DONERT, K. et al. (Eds.). *Learning with Geoinformation III*. Heidelberg: Ed. Wichmann, pp. 96-105.

Mínguez, M.C. (2011), Enseñar Geografía a través del análisis de imágenes con la ayuda del campus virtual y la pizarra digital interactive, in Delgado, J.L.; Lázaro, M.L. y Marrón, M.J. (Eds). 2011. *Aportaciones de la Geografía para aprender a lo largo de la vida.* Málaga: Grupo de Didáctica de la Geografía (A.G.E.) y Universidad de Málaga. (Obra en CD).

Schultz, R.B., Kerski, J. J. and Patterson, T. (2008), The use of virtual globes as a spatial teaching tool with suggestions for metadata standards. *Journal of Geography* 107: 27-34.

UNESCO. (2008), *ICT Competencies Standards for Teachers.* <http://cst.unescoci.org/sites/projects/cst/The%20Standards/ICT-CST-Implementation%20Guidelines.pdf> (07/2012)

Valenzuela, M., Molla, M; and Lázaro, M. L. (2004), Geography in Spain. In: *Belgeo*, 1, Special Issue 30th International Geographical Congress, pp. 145-161.

PART TWO:

NATIONAL AND CASE STUDIES

CHAPTER SIX

PAIKKAOPPI: A WEB BASED LEARNING ENVIRONMENT FOR FINNISH SCHOOLS

LEA HOUTSONEN, SANNA MÄKI, JUHA RIIHELÄ, TUULI TOIVONEN AND JUKKA TULIVUORI

GIS in the Finnish school curriculum: Pedagogical viewpoints

Nowadays, teaching using GIS in schools takes place in several countries because of its close links to everyday life in connection with work or free time activities (Milson et al. 2011). However, the status of GIS in the school curriculum varies greatly around the world. In Finland, under education legislation, the teaching of GIS should be provided in all upper secondary schools in connection with the optional specialised course entitled "Regional Studies" (Finnish National Board of Education 2004). According to the national core curriculum, the content of GIS teaching should comprise the basics of Geographical Information Systems and their applications, as well as examples of processing, interpretation, and visualisation of geographical source material at different levels using GIS software. The aim is to use the teaching of GIS in schools to educate young citizens with the skills necessary to participate in the information society in both work and free time uses. The use of GIS in education should also promote sustainable development and encourage students to actively take part in planning and developing their own environment. In 2004, the Finnish national core curriculum introduced cross-curricular themes such as active learning, technology and society, active citizenship and media skills, and GIS education is connected to each of them.

In the Finnish national curriculum for geography, applying GIS as a tool for making individual regional analyses became reality, at least on paper. In practice, however, many teachers have been puzzled by the world of different GIS software, data formats and various technical issues. Data products offered by Finnish authorities tended to be expensive and buying them from different sources was considered too time-consuming. The idea of using GIS for active learning was widely recognised, but practical realisation seemed impossible to achieve in most schools. As a result, students were in danger of being treated unequally, as the use of GIS depended on the technical interests of the teachers, the school budget and on the data policy of the municipality where the school was located.

In the period 2012 to 2016, Finnish schools are on the threshold of major curriculum reform. The National Board of Education has been tasked with revising the national core curriculum, to which municipalities and then schools will then revise their own local curricula. The National Board of Education will then be responsible for guiding and supporting the implementation of the national core curriculum, which means it has the mission of completing the process of drawing up municipal and school-based curricula. The coming period of reform opens new opportunities to examine the state and content of GIS education in Finnish schools.

Concerning curriculum planning, international trends in education seem to have shifted from traditional teacher-centred approaches towards a student-centred paradigm, i.e. the focus is not only on teaching, but also on the skills the students are expected to have at the completion of courses. This is commonly referred to as an outcome-based process. In recent years, attempts have been made to revisit and revise Bloom´s taxonomy of educational objectives, but the original works of Bloom are still the most widely quoted (Kennedy, 2012). Bloom proposed the cognitive domain should be composed of six successive levels arranged in a hierarchy:

1. Knowledge
2. Comprehension
3. Application
4. Analysis
5. Synthesis
6. Evaluation

Bloom’s taxonomy has been a very useful tool when defining learning outcomes in geography and for GIS education. Of particular significance is the nature of the contribution that teaching in GIS can make to the development of pupils’ critical faculties and especially their capabilities in terms of analysis, synthesis and evaluation. Fitzpatrick and Maguire (2001) consider this is linked to the development of their logical-mathematical, linguistic, spatial, and interpersonal intelligence, where "logical-mathematical intelligence includes numeracy and technological capacity, linguistic intelligence includes literacy and graphicacy, spatial intelligence includes map literacy, and interpersonal intelligence focuses on communication".

The increasing demand for virtual data and information on the environment, as well as the adoption of ICT and its importance in the society, means that almost all citizens will need to have the ability to handle spatial data in the future. Teaching of GIS can develop students' logical thinking and problem-solving abilities (Green, 2001). Learning with GIS enhances students′ information processing skills, reasoning skills, questioning skills, creative thinking and evaluation skills.

In teaching geography, the use of GIS can improve students' understanding of spatial concepts. However, more research is needed to understand whether increasing spatial understanding by teaching with GIS differs from increasing it through teaching of conventional cartography (Bednarz, 2004).

Motivation is one of the most important factors affecting students′ attitudes towards learning geography. The enthusiasm of students can be increased when using enquiry approaches with GIS within normal teaching processes (Artvinli, 2010). These perspectives are also important when considering the inclusion of GIS education in the school curriculum.

The use of GIS in geography teaching particularly as a means of increasing spatial understanding can be justified on the following grounds: geography as a science is devoted to the study of spatial order and spatial dependence relations, and its central concepts include spatial differentiation, landscape, environment (man-nature interactions), distributions, geometric features (networks), localisation, the determination of location, etc. Maps have traditionally been used to study spatial dependence relations, and "overlay transparencies" is the classical method for identifying these. GIS education thus supports students′ geographical skills by improving their spatial thinking.

Artvinli (ibid) stated that there are four levels of GIS in education:

1. Teaching about GIS (teacher-centred)
2. Teaching with GIS (teacher-centred)
3. Learning with GIS (teacher-centred)
4. Research with GIS (student-centred)

He suggested GIS applications in secondary school geography should be planned at the level of "learning with GIS" and "researching with GIS".

Teaching with GIS is in line with many of the basic principles of constructivism. The students are active processors of information who are engaged alongside other students, their teachers, and their environment in handling and interpreting data, which reflects situations in the real world on the basis of previously acquired cognitive structures.

Development of the PaikkaOppi learning environment

The PaikkaOppi project started in Finland in 2007 as a pilot with the aim of creating a web-based GIS platform that combines spatial data from various topics with a pedagogical web-mapping tool. The project was led by the Lounaispaikka network, and coordinated by the Regional Council of Southwest Finland. Technical expertise was provided by a private GIS company, Arbonaut, and the Geodetic Survey of Finland. Three pilot schools in the cities of Joensuu and Turku played a crucial role in commenting and implementing the pedagogical part. The universities of Helsinki and Turku provided further pedagogical expertise. The project was financially supported and directed by the Finnish National Board of Education.

The mapping application, which formed the central part of the learning environment, was based on several criteria. The coverage of the data sets provided had to be for the whole of Finland to ensure equality among the students. The data had to be detailed enough for geographic activities to take place in the school surroundings. The content had to be thematically extensive to allow multiple inspection angles of the environment. In addition, the service had to support active and collaborative learning. Hence individual mapping exercises, sharing personal data with other users, and reporting to the teachers had to be enabled. In addition, it was important that the service was usable, easily accessible through standard web browser, and technically stable for teachers to adopt it.

The mapping application produced during the project allows the visual analyses of the most detailed environmental data sets in Finland, from different topographic maps and aerial photos to information on soils, fluvial system, infrastructure, and population. For some areas, historical maps are also included. The majority of data is available from scale 1:20000 onwards, covering the entire country. The data sets are up-to-date as the application uses WMS and WFS services offered by data providers to link data sets directly to the map service.

Individual point and tracking data can be entered directly from a GPS device. This data may be classified in different attribute classes represented by different symbology. Individual interpretations can be drawn on top of other data layers as points, lines and polygons with attributes. The application also included a tool used to visualise municipality level population data with interactive thematic mapper.

Sharing reports with others is supported by PaikkaOppi design. The map output can be saved as a workspace and opened later in the map service, or it can be stored as an image and exported to PaikkaOppi's own wiki-based workbook. The workbook provides each user a personal space to which content can be added. Students can use the space to save documents, assignments or simply as a library to store their map output. The content saved to the workbook can be shared with all users, also with external users since browsing the workbook does not require registration.

The workbook also included pedagogical content created during the project. Pedagogical content was designed to vary in terms of topic and scale, and to give students the possibility to use the service in a variety of settings — in school, at home, independently, or in a group. Assignments varied from basic understanding of cartography to applying GIS data sets in understanding physical, cultural and economic characteristics locally and regionally. The assignments were designed so that users required no previous GIS experience other than being able to navigate in the map window.

The development of PaikkaOppi has been a technological experiment. The service has evolved in line with web mapping techniques, tools, and standards. The service has been developed using open source tools and taking advantage of recent web mapping and data sharing standards. Content building has benefitted from the implementation of the INSPIRE directive in Finland, as the number of WMS and WFS services has grown during the years of development.

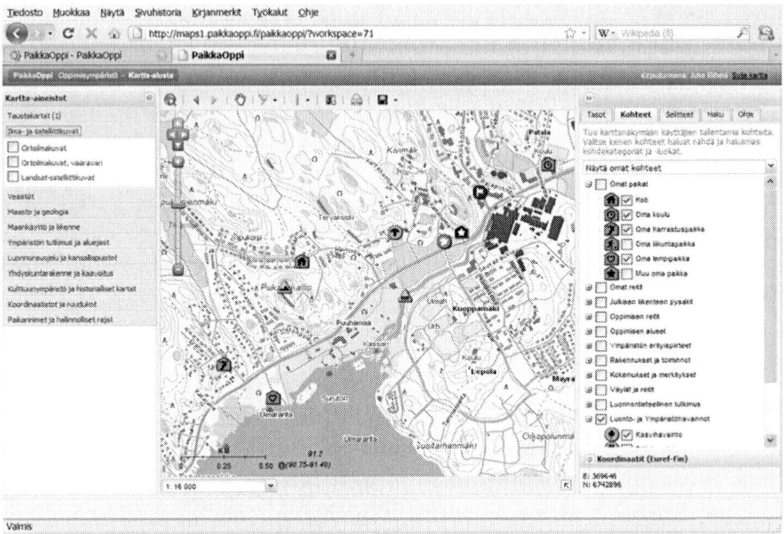

Figure 6.1. PaikkaOppi desktop view. (Photo: Juha Riihelä).

Service in use

The first courses using PaikkaOppi were arranged in the pilot schools in the autumn of 2008. Since then, the platform has been widely used in geography courses. The users range from students in lower grades of basic education to those on special courses of GEOGRAPHY in the upper secondary schools. Recently, the service has also raised interest at university level.

In geography courses, students have used PaikkaOppi to study the basic geography of Finland. This is facilitated by the broad content of the service and by personal observations made with GPS devices. The basics of GIS become familiar while studying the environment: students work with different data sets, browse through them and their metadata, as well as compare them by overlays in the map service.

PaikkaOppi was also designed to be useful in disciplines other than geography. In particular, biology and social sciences, for example history, were seen to benefit from the GIS tools and datasets provided by the platform. An example of a cross-disciplinary activity is a landscape history course arranged in the city of Turku during the spring of 2010. The teachers of geography, history and landscape studies planned the course jointly. The aim was to study changes in the landscape, and in particular physical features and man-made structures. PaikkaOppi's map service was used to study and compare original maps dating from the 18th century to contemporary maps and aerial photographs. Data sets were overlaid in the map service, and basic visual analyses could be executed.

Figure 6.2. Students on a field trip in Halikko, Finland. (Photo: Juha Riihelä).

What do Finnish teachers think about PaikkaOppi?

It is always important to know how teachers use ICT in their lessons. Based on a geographical study made during summer 2011, it can be concluded that Finnish geography teachers use information and communication technology in teaching in a very diverse way (Tulivuori, 2011). However, the uses depend mostly on the resources available that in turn depends on these local education providers. The study had two parts, one quantitative and the other qualitative part.

The quantitative research showed that information and communication technology has been mostly used for teacher-led teaching, mainly to illustrate and clarify the various key phenomena of geography, rather than students having access to do something themselves. Geography teachers in their own work used different GIS software and PaikkaOppi surprisingly infrequently. The main obstacles to their use of GIS software or PaikkaOppi was, again, lack of resources, but also a deficiency in the in-service training available. The teachers clearly felt they had not received adequate support for the use of ICT and GIS software during their studies, in addition, the in-service training offered by education providers was not considered sufficient, even if it took place during normal working hours.

In the section concerning geography teaching in upper secondary school, the national core curriculum includes three courses in which information and communication technology can be quite easily used to improve learning. PaikkaOppi was generally seen as a positive phenomenon for teaching and learning these areas, and teachers also considered teaching and learning objectives to be quite easily achievable when using it. In many cases however, the informants clearly drew attention to the fact that many of the objectives were quite challenging and did not necessarily take place in the everyday life of school. In addition, the content of geography courses was often criticised as being too broad and demanding without focus.

The results of the qualitative part of the study were similar to those of the quantitative part: geography teachers said they used information and communication technology in their work as much as they could with the given resources. Also, most headmasters and teachers of other subjects had a positive attitude towards the use of ICT for teaching purposes, but they were not necessarily as advanced as geography teachers in their own use. It was clear that peer support was important for geography teachers to adopt new innovations in their teaching. New ideas were also drawn from in-service training, but this that was, however, said to be inadequate. The respondents suggested special attention should also be given to the students' level of knowledge. Nowadays students' computing skills, for example, may vary in the beginning of upper secondary school.

When asked about PaikkaOppi in particular, the following results were obtained:

Firstly, how the use of PaikkaOppi supported individuality in learning was asked (in terms of different ways to build knowledge and assimilate new information). The majority of survey participants (90.9%) thought that PaikkaOppi supported individuality in learning.

> "*It enables differentiation and the implementation of individual learning paths.*" (Male, 36-50 years),

or

> "*Those who have interest in information technology are more likely to become motivated.*" (Female, 26-35 years).

Meeting this objective also supports a constructivist learning theory. The learner no longer only possesses knowledge, but also produces and builds knowledge and personal beliefs. Similarly, this objective is strongly related to introducing geographic information systems. (Ilomäki, 2008, 20; Bednarz, 2001 3-7.)

Secondly, it was asked whether PaikkaOppi supports community (collective) learning (when knowledge is produced and topics are discussed with others). The respondents' answers were not quite as straightforward but 84.8% considered this support successful. Many respondents, however, gave good examples of the increase of community learning while using PaikkaOppi.

> "*Let's work together, for example, first in terrain GPS points, and then examine the self-made maps together*." (Female, 26-35 years),

> "*Individual output is put together as a final report which will be worked on together*." (Female, 51 years),

or

> "*Combining collaborative learning with pair work, and, of course, when there is previous course work to continue with*." (Female, 51 years).

According to constructivist learning theory, collaborative learning, during which the learning path is built together, is also important (Lomami 2008).

The third question asked whether PaikkaOppi supports networking between students, teachers, school and surrounding communities. Slightly less than a quarter (24.2 per cent) of the respondents considered this objective achievable.

> "*We have not been working collaboratively in PaikkaOppi*." (Female, 26-35 years).

On the other hand, it was felt that the teacher's guidance was needed if collaboration was to take place.

> "*It does support networking if the group agrees to work together, free-riding in the Internet is easier when the teacher is not watching all the time*." (Female, 36-50 years).

In addition, the following opinion appeared in the answers:

"*The teacher has to be to where the students are (online), but the use of ICT itself does not guarantee networking, etc., but it requires that the teacher wants to promote these things*." (Female, 26-35 years).

According to Ilomäki (2008), information and communication technology plays, and continues to play, an important role in networking within a school, as well as within its immediate surroundings. With modern means of networking available, it can be done easily and in real time.

Fourthly, it was asked whether PaikkaOppi supports teacher-student relations, data ownership, and the democratic transformation of knowledge production. This question had the highest number of negative answers. Almost 40 per cent of respondents felt that the teacher-student relationship does not change, with the use of the PaikkaOppi system. On the other hand, some of them felt that the teacher-student relationship did change.

"*The teacher can give the student feedback without other people interfering with it*." (Female, 51 years),

or

"*Students can produce important information from their own point of view*." (Male, 36-50 years).

However, such a change in the ways schools and teachers approach learning and teaching does not happen overnight. The PaikkaOppi learning environment, however, certainly contributes to fracturing the traditional boundaries of a school.

Figure 6.3. Students on a field trip in Lemmenjoki, Northern Finland, using GPS devices. (Photo: Jukka Tulivuori).

The Finnish national core curriculum (2012-2016)

The revised national core curriculum for Finnish schools will be put into practice in August 2016. The new curriculum will underline the skills of the students and investigative learning will be one way to acquire and produce knowledge. Students will be guided in the use of information and communication technology and services, with the objective to strengthen the use of GIS education in the new curriculum.

The new core curriculum is meant to include tools to promote learning and integrate applications of information and communication technology that contribute to the management, analysis and visual presentation of spatial data. GIS teaching leads students to make their own geographical observations instead of reading about those made by others. It introduces them to many topics that lie at the very heart of geography. They can improve their cartographic skills, learn to interpret natural and cultural landscapes, and attempt to perceive interaction relations between phenomena. GIS teaching also allows them to develop their skills in influencing decisions made in the society, and opens up opportunities for them to take part actively in developing their own community. The pedagogical development of GIS and its teaching remains one of the major challenges for geographical education in schools (Houtsonen, 2003). PaikkaOppi will remain an important tool for teachers to use in the future.

Conclusions

In addition, the Finnish Matriculation Examination will be digitised in 2016. Geography will be one of the pilot subjects in this process, and PaikkaOppi will also be needed as part of the new digital exam.

The pilot phase of PaikkaOppi began with three schools in the Joensuu and Turku regions. Since 2008, PaikkaOppi has been actively used in many schools around Finland. Today, there are more than 900 registered teachers and the number of monthly users has increased to around 5,000. However, the service still has many things that need to be developed and improved.

In future, PaikkaOppi will hopefully be maintained together by the Finnish National Board of Education and the National Land Survey of Finland. This is crucial, as there will a big need for services like PaikkaOppi once the revised national core curriculum will be put into practice in 2016. Teachers also widely support the use of PaikkaOppi in teaching and of course in student's learning, too. In addition, in 2016, the plan is for the Finnish Matriculation Examination to become digital.

Geography will be one of the pilot subjects in this process and PaikkaOppi development will be needed if it is to be an important part of the new public examinations.

References

Artvinli, E. (2010). The Contribution of Geographic Information Systems (GIS) to Geography Education and Secondary School Students' Attitudes Related to GIS. *Educational Sciences: Theory and Practice*, 10(3), pp. 1277-1292.

Bednarz, S. (2001). Thinking Spatially: Incorporating Geographic Information Science in Pre and Post Secondary Education. Houtsonen, L. & Tammilehto, M.: *Innovative Practises in Geographical Education. Proceedings*, 3–7. Helsinki, Finland, August 6th- 10th. IGU, Commission on Geographical Education.

—. (2004). Geographic information systems: A tool to support Geography and environmental education. *GeoJournal* 60(2), pp.191-199.

Finnish National Board of Education (2004). *National Core Curriculum for Upper Secondary Schools* 2003. Helsinki. 261 p.

Fitzpatrick, C. & Maguire, D.J. (2001). GIS in schools: Infrastructure, methodology and role. In Green, D.R. (ed.) *GIS: A Sourcebook for Schools*. Taylor & Francis, London, pp. 62-72.

Green, D.R. (2001). GIS in School Education: You don't necessarily need a microcomputer. In Green, D.R. (ed.) *GIS: A Sourcebook for Schools*. Taylor & Francis, London, pp.34-61.

Houtsonen, L. (2003). Maximising the use of communication technologies in geographical education. In Gerber, R. (ed.) *International Handbook on Geographical Education*. Kluwer Academic Publishers, Dordrecht, pp.47-63.

Ilomäki, L. (2008). *The effects of ICT on school: teachers' and students' perspectives*. Painosalama, Turku. 78 p.

Kennedy, D. (2012). Working Towards a Common Language for PROFILES Modules. In Bolte, C., Holbrook, J. Rauch, F. (eds.) *Inquiry-based Science Education in Europe: Reflections from the PROFILES Project*. 1st International PROFILES Conference 24th-26th September 2012. Freie Universität Berlin, pp. 185-189.

Milson, A. J., Demirci, A. & Kerski, J. J. & (eds.) (2011). *Teaching and Learning with GIS in Secondary Schools*. Springer, Netherlands. 225 p.

Tulivuori, J. (2011). *The Use of ICT in Teaching of Geography*. Master's Thesis (Unpublished). University of Turku, Department of Geography and Geology. 64 p.

CHAPTER SEVEN

INTRODUCING SPATIAL LITERACY CONCEPTS TO MIDDLE SCHOOL STUDENTS IN ALBANIA: A CASE STUDY FROM THE UNIVERSITY OF FLORIDA

JUNA PAPAJORGJI

Hapu, hapu, errësirë! Pa jakë tëhu, o drive!
[Clear out, you darkness! And come in, you light!]
—Naim Frashëri, 1887[1]

Introduction

As part of the development of an international curriculum providing integrated teaching of spatial literacy and urban sustainability concepts to secondary education students, the Urban and Regional Planning department at the University of Florida, USA has developed a prototype course that uses Geographic Information Systems (GIS) to teach Albanian middle school children about environmental and sustainability subjects. The course, titled "Albanian Youth for Environmental Education" (AYFEED), was initially taught in June 2012 to a mixed age group of students from the randomly selected public middle school "Dëshmorët e Lirisë" in Tirana (the capital of Albania). The entire course and related data, including the products created by the students, are openly available online, from several websites, through its main portal at: http://ayfeed .wordpress.com/.

[1] An excerpt from the poem "Korça", dedicated to the opening of the first school in the Albanian language—the result of long struggles for the preservation of national identity.

Several partner organizations provided significant resources to this effort. They include the GISCorps,[2] which provides worldwide volunteer GIS services to less advantaged communities; the Green Mapping System,[3] which engages worldwide communities in mapping green living, nature, and cultural resources; the Mediterranean Association for the Protection of the Sea Turtles (MEDASSET),[4] and the My Community, Our Earth,[5] an

---

[2] GISCorps (http://www.giscorps.org) is an international programme of the Urban and Regional Information Systems Association (URISA), which is headquartered in Chicago, Illinois. GISCorps coordinates short-term volunteer GIS services to needy communities worldwide. Its partners include UN agencies (UNDP, UNOSAT, WFP, UNJLC, WHO, UN-Spider), Emergency Operation Centres in the United States, Amnesty International, the World Vision, the Global Spatial Data Infrastructure (GSDI), Engineers without Borders, the OpenStreetMap, and the Australian and American Red Cross organizations among others. Since its inception in October 2003, the Corps has attracted over 3,000 volunteers, from 95 countries. Up to June 2013, GISCorps had deployed 422 volunteers to 122 missions, in 47 countries. These volunteers have contributed to missions in Africa, Asia, Eastern Europe, Latin America, and the United States. They have expertise in: remote sensing, GIS analysis, GIS database development, Web-GIS application development, GPS usage and processing, GIS teaching, and expert crowdsourcing.

[3] The Green Mapping System (http://www.greenmap.org), headquartered in New York City, New York, was founded by Wendy Brawer in 1995 to address the need for greener, healthier cities. It is based on a product-service model that combines a universal iconography, adaptable tools (online, and off line), and local leadership. It thus offers access to a global collection of sustainable maps and their map-making tools. Green Mappers include youth, designers, social entrepreneurs, non-for-profit organizations, universities, governmental organizations, and tourism agencies. Their cumulative effort has come to be known as the Green Mapping Movement, and to date it has reached at least 625 cities, towns, and villages, located in 55 countries across the globe. The focus of the programme is equally divided between the process of map creation, and the outcome map product. Each locally-led Green Mapping project has a unique way of involving people of all ages in discussing, assessing, and highlighting green living resources and sites of natural, social, and cultural value. Green Mappers build skill sets as they organise, plan, design, and promote their maps or their interactive workshops and tours.

[4] MEDASSET (http://www.medasset.org) is an international environmental non-governmental, not-for-profit organisation, registered in the UK and in Greece. It was founded in 1988 (with roots going back to 1983). Its mission is to gain public support for establishing sea turtles as a flagship species for marine and coastal biotopes conservation needs throughout the Mediterranean region, and to ensure that these needs become central to national and international policy. MEDASSET's research and conservation projects focus on areas where projects have not been carried out before, or where little or no commitment to sea turtle conservation exists. Since 1988, over 7,800 km of coastline has been surveyed,

international program of the American Geographers Association. The remainder of this chapter is a brief report about this effort.

Background

Albania (or Shqipëria) is a Mediterranean country with a transition economy. It is also a land whose inhabitants greet each other daily with the expression "may your life be extended".[6] Albania is a predominantly mountainous country, with a third of its territory covered by forests, and a third of its boundary lined by water (the Adriatic and Ionian seas). Although one of the smallest countries in Europe, Albania retains a rich biological and landscape diversity which spans a wide range of unique flora and fauna (NEA, 1999). But Albania (which until the 1990s lacked legislative mandates for environmental protection) is, by European standards, also an economically poor country. Ensuring the protection of its rich ecological systems, or expressly educating its youth in these matters, has been seen as a top priority by its post-communist governments.

Following the collapse of communism in 1991, and the country's growing interaction with international organizations and their financial support, the past two decades have introduced improvements in the quality and structure of Albanian education. However, significant strides are yet to be made in many areas, including two critical ones: environmental science and computerised technology. Most schools in Albania, and especially schools in rural areas, have yet to provide students with access to technology, and do not include computers in the daily routines of teachers and students.

---

from Sardinia to the NE Ægean and from the Ionian Sea to the shores of Egypt and Libya. Their main activities include: research and conservation, education, training, capacity building, awareness raising, contribution to policymaking processes, and lobbying and advocacy.

[5] My Community, Our Earth (MyCOE) (http://www.mycoe.org), is a programe of the Association of the American Geographers (AAG). It was created as a private-public partnership in the year 2002, preceding the World Summit for Sustainable Development (WSSD) in Johannesburg, South Africa. Its goal is to address sustainable development issues, including biodiversity, climate change, poverty eradication, fresh water supply, and urbanisation. The programme provides geographic perspectives, learning resources, and technological tools in support of youth that engages in their local communities around global sustainability themes.

[6] The Albanian word *tungjatjeta* is a widely used greeting which complements other greetings that reference the time of the day.

It is important to note that although Albanians have had a continuous presence in the Balkan Peninsula since prehistoric times, and although they speak an Indo-European language, which constitutes one of its ten original branches (Campanile et al, 2005), Albania is also a very young nation-state. Albania's long history of oppression and invasion (Zoi, 2000), has conditioned a very young national educational framework. As such, the country is still grasping with the ideal shape of its educational systems and institutions. After a fierce struggle for national survival, it was only in 1844 that the country's first ABC book[7] in the national language was published, and it was only in 1860 that its first newspaper[8] was published. Its first school in the national tongue was opened in 1887, and its first general education high school[9] was opened in 1917. The literacy rate of the country, which is now 98.7 per cent (UNDP, 2011), remained below 10 per cent until 1945 (Cook, 2001), decades after its independence (1912).

Figure 7.1. Sole surviving photograph from the opening of the first school in the Albanian language. Korçë, 1887. (Photographer, Kristo Panajot Shuli.)

[7] The first ABC book in the Albanian language, called Evëtar, and which initially contained eight pages, was authored by Naum Veqilharxhi from the Albanian Diaspora in Romania, and was first published in Bucharest, Romania, in 1844 (Hyseni, 2012).

[8] The first ever Albanian newspaper is considered to be "Shqiptari i Italisë", published in the Albanian diaspora in Italy by Jeronim De Rada, in 1860.

[9] The first general high school in Albania, the well-known "*Lycée de Korçë*", was opened under the French protectorate on October 25, 1917, in Korçë. Its first cohort had 36 students. Of them, only nine graduated. Of these, two were this author's uncles (Bino, 1999).

In this context, and to support the education of Albanian youth in natural resource protection and in geospatial literacy simultaneously, the University of Florida, USA, proposed a short-term curriculum entitled Albanian Youth for Environmental Education (AYFEED). AYFEED is an environmental educational pilot workshop that was first presented to public middle school students in Tirana, the capital of Albania. The workshop introduced environmental and sustainability concepts in the context of Albania's history, geography, environment, and culture by means of one of the fastest growing disciplines (Batty, 2012), Geographic Information Systems or Geographical Information Science (GIS). GIS, a modern model consisting of an information technology framework, which is increasingly becoming a mainstream modus operandi. It provides an attractive and visual way to monitor and analyse the environment and a powerful setting for communicating results via maps. While students learn the valuable skills of GIS, they also discover the importance of interrelated systems within the environment. This, in turn, gives them a sense of responsibility and independence that can stimulate involved, independent thinking, and hence, conscious citizenship.

The AYFEED Project Directive

Responding to calls from Albanian government institutions for support and cooperation from non-governmental organizations in helping the country progress towards a more educated and open society, two partner organizations offered partial funding to the University of Florida's proposal for the AYFEED project. These partner organizations, with the encompassing acronym MMIF,[10] consist of the Martin and Mirash Ivanaj Foundation (a not-for-profit organization based in New York City, United States), and the M. & M. Ivanaj Foundation Institute (a not-for-profit organization based in Tirana, Albania). The mission of MMIF "is to help

[10] The MMIF was first founded in New York City in 1995, by Drita Ivanaj. It is a memorial to two Albanian brothers: Martin Ivanaj (1888-1940), and Mirash Ivanaj (1891-1953). Mirash Ivanaj is a respected personality in the history of the Albanian education, known for his role in standardizing, consolidating, and expanding the national public education system in Albania during his tenure as Minister of Education in the pre-communist government. He is remembered for the law that made elementary education compulsory nationwide. He died in prison as one of the innocent victims of the communist purges of patriots and intellectuals. His brother, Martin Ivanaj, was an attorney general in the pre-communist government of Albania. They were educated at the University of Rome, in Italy, and were both writers, scholars, and prominent intellectuals (Gogaj, 2004).

the young generation of Albania advance their education in the preservation of freedom and independence of their country…" by "…promoting and encouraging culture and education" (Ivanaj, 2008).

In addition to providing partial funding, the MMIF also engaged in securing and selecting the school, in supporting in-country related problems of infrastructure administration and logistics coordination, and in developing in-country support and partnerships for the facilitation of the implementation of the project. These partnerships included the Albanian Ministry of Science and Education, the National Albanian American Council, the Regional Environmental Center for Central and Eastern Europe, the Albanian Minister for Innovation, Information, and Communication Technology (also a parliamentarian), and the first United States Ambassador to Albania.[11]

Objectives

The short-term objective of this project was the development of a prototype course that uses Geographic Information Systems (GIS) to teach Albanian middle school children about environmental and sustainability subjects. The longer-term objective of this project was its positioning as a prototype case study towards the development of a broader short-term international curriculum that would provide integrated teaching of spatial literacy and urban planning concepts to young students. A related strategic objective was the preliminary establishment of a trans-organizational and trans-national collaborative institutional network as an ongoing sustainable resource sharing and knowledge exchange framework.

The specific technological objectives were as follows:

- use freely available software and data;
- use both online and off line resources;
- use a combination of local, country, regional, and global data;
- place all course products in the public domain.

Specific pedagogical objectives were to:

- teach with GIS rather than about GIS;
- teach in English but maintaining a strong local context;
- integrate technology seamlessly across many media forms;

[11] William Ryerson was the first U.S. ambassador to Albania after the two countries restored diplomatic relations in 1991, which had been interrupted in 1946.

- integrate topics seamlessly across disciplines;
- connect student's personal experience to the larger world;
- create a mixed age classroom rather than a single age classroom;
- balance student's role as consumers of knowledge versus that as creators of it;
- create a network classroom structure versus a hierarchical one.

Method

The work for this project included seven distinct components. They were not conducted in a chronological order, and regularly overlapped in time. Hence, these components are listed below not necessarily in the order in which they were completed.

- exploration and definitions
- data collection and database development
- development of the course
- deployment of the results
- teaching of the course
- coordination and logistics planning with hosts in Albania
- partnering with trans-national organizations and programmes

The "exploration and definitions" phase was employed to review the previous work conducted in this area, and to appraise the current education landscape in Albania. The "data collection and database development" phase dealt with exploring, gathering, judging, and manipulating the necessary data. During the "development of the course" phase, the book lessons and the related course exercises were designed and developed, and then thoroughly tested, and retested in terms of the accuracy of the data, and the projects. During the "deployment of the results" phase a website was developed that would serve as a central hub for hosting the course products, and the links to its auxiliary resources. The book-lesson publications were developed and their accompanying data and software, and the course installation was structured and tested with its related step-by-step documentation. During the "teaching of the course" stage, several days were spent in Tirana, Albania, where the lab was prepared, teaching took place, and post teaching evaluation activities were undertaken. The next two components related to in-country logistics and strategic partnerships, did not fit into one explicit timeframe, although they were very time and attention demanding, and although they were very distinct activities. These two components constantly and dynamically permeated the entire length of the project.

Exploration and Definitions

Research into the most pressing contemporary environmental issues in Albania and the state of environmental education for our designated age group was conducted in this phase. The opinion of Albanian environmental scientists was also sought on the content of the curriculum, via informal personal and professional relationships. Extensive research was also conducted for reviewing prior efforts from the United States higher education community with teaching GIS to secondary school students. Further research was also conducted in this phase to determine the most appropriate GIS software for use. As acquiring proprietary software would have increased the cost of implementation and re-implementation of the project, and it would have subsequently negatively affected its long-term sustainability, we set out to design this course based on free software. Our research therefore examined two avenues: the adequacy of using Free and Open Source Software, and that of using free, but not open, shareware software. We determined that given the near fifty per cent rate of Internet penetration in Albania (The World Bank, 2011), an equal combination of online and offline software was a sensible, balanced approach. In the end, we decided to use a combination of the ArcGIS Online Map Services and the ArcGIS Explorer from the Environmental Systems Research Institute (ESRI), and the free version of the Google Earth software from Google, Inc.

Data Collection and Database Development

We dedicated this stage to compiling a library of environmental and geographic databases for Albania. We set out to rely on data that was in the public domain, and we expended significant efforts in standardizing, translating, and enhancing the data that we found. We afterwards added new descriptions and information to these databases, which we tailored to our designated age group and to our specific exercises for the course. In addition, we also engaged in compiling another library of environmental and geographic databases from the world and the region, and which also had particular significance to the topics of our course. A similar example is the land-to-water cosmopolitan itinerary of the sea turtles, a protected species worldwide, and which has an important breeding presence in the bay of Patoku (White et al, 2008), along the Adriatic coast of Albania. In sum, finding public domain GIS data about Albania turned out to be a very onerous task. Very little or close to nothing seemed to exist. The sparse data that we found was in close to sub-standard conditions, and it required a great deal of effort for it to become meaningful and useable. Outreach efforts to public data holding agencies in the country did not yield any successful results.

Development of the Course

At this stage, a draft GIS curriculum and its corresponding geographic data library were first prepared for the course. This GIS draft was followed by a draft environmental curriculum. Both of these drafts were then merged into one curriculum, and were tested and re-tested for (a) the age group, (b) cultural appropriateness, and (c) second language clarity. Afterwards, this new curriculum and the two corresponding geographic data libraries were integrated, and structured into one single product to be used by the students and/or teachers.

The curriculum includes six stand-alone book lessons, which were designed with step-by-step instructions and with screen captures, and which are bundled with their corresponding data and software projects. Each of these book-lessons was approximately fifteen pages, and was paced for a young student with limited proficiency in English. An example of one exercise, from one of the lessons, is shown in Figure 7-3 below.

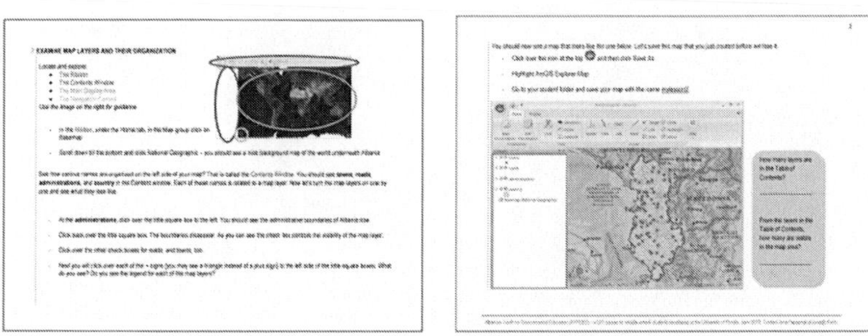

Figure 7.2. Example from Lesson 6 titled "My waste and the sea turtle."

A day long final project was designed for the fourth day of the course. In an effort to put into practice the knowledge acquired during the first three days of the course, and to also encourage students' creation of authentic work based on their own knowledge of their surroundings, the final project asked the students to work in teams for composing a Green

Map for Tirana (its first). Students were asked to create two alternatives for Tirana's Green Map. One map showing the location of green resources at present, the other map showing the location of proposed green resources as recommended by them. During the morning session students were asked to discuss and reach a consensus and to collaboratively hand-draw their maps on paper, using globally standardised icons and symbols. During the afternoon session they were asked to import and or translate the paper products into a GIS framework. The day ended with team presentations given by students (in English).

This approach places a premium on using technology-based collaborative ways for creating new content that is generated by students themselves in support of their own learning process. It positions students in the role of producers of the primary source of knowledge as opposed to their more traditional role as consumers. In the traditional role, students are passive "consumers of knowledge," and the content that they generate is mostly used by their teachers "for their evaluation" (Herrera et al, 2012).

The curriculum also included a course outline, a course schedule and syllabus, a reference list of resources in GIS and environmental science for this age group, course evaluation forms, an online student survey, course certificates, and other supporting documents for the course and its deployment and installation. All are openly and freely available online.

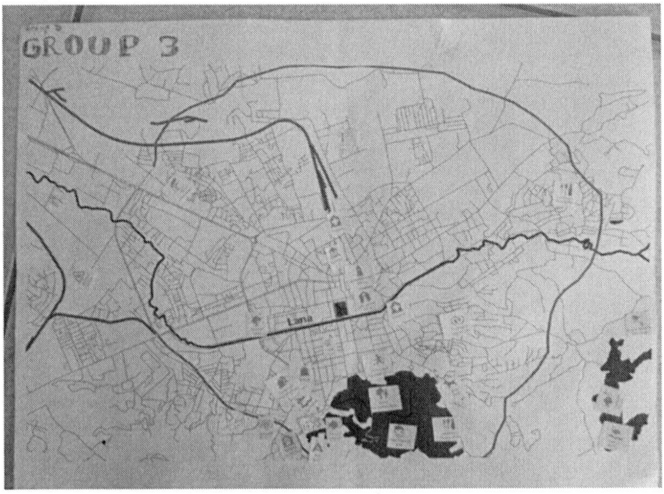

Figure 7.3. Example of a hand drawn green map.

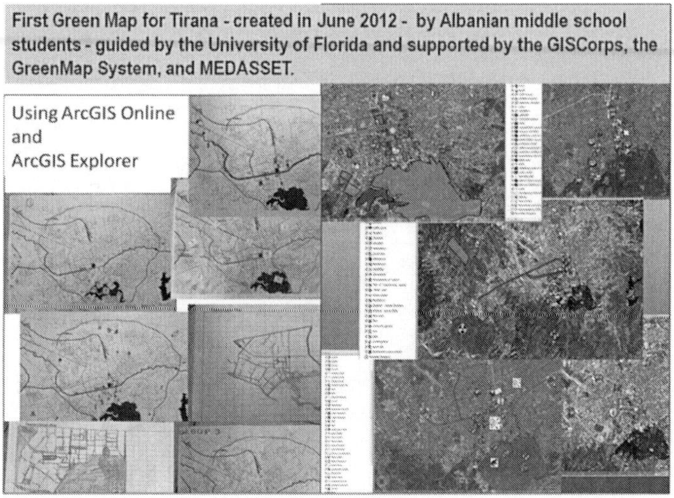

Figure 7.4. Green maps side by side - hand made and electronic.

Deployment of the Results

In this stage, we engaged in the bundling and distribution of the course. Our goal was to design the product for future use and reuse by anyone with minimum computer knowledge. The data, the software, the project applications, the course curriculum, and the workbooks, were assembled into one single integrated medium. Based on the circumstances, this medium could be received as a DVD, as an online arrangement, or in any other contemporary digital media distribution form. We placed all of the products in the public domain.

As shown in Figure 7-6 below, we also developed a website that hosts the entire course materials, software, and related data and products: http://ayfeed.wordpress.com/. The entire course and related data and products created by the students during the workshop, are available on ESRI's web site. Lessons and data are at: http://edcommunity.esri.com/arclessons/arclessons.cfm. Students' final work is at: ESRI ArcGIS Online. The students Green Maps, and the letter sent to the Mayor of Tirana, are published at the Green Mapping System, and at the My Community, Our Earth.

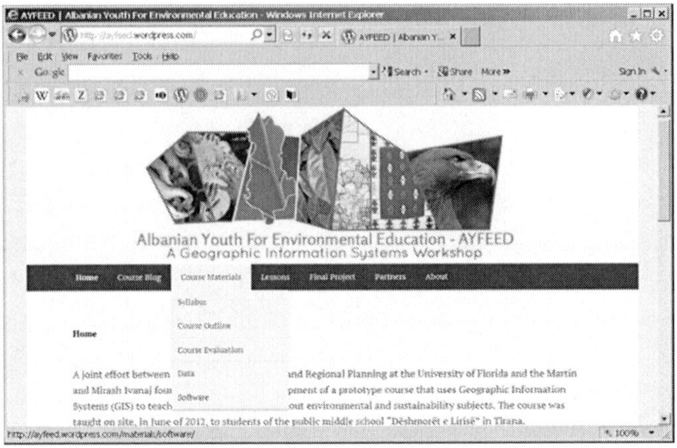

Fig. 7.5. Open course availability from the web.

Teaching of the Course

This phase lasted seven days. During the first two days, three teachers were based in the school lab for nine to ten hours each day, intensively preparing it for the workshop. This preparation included installation of parent and other software, installation of the course data and projects, testing, overcoming issues of technological incompatibilities, and braving unanticipated obstacles of network and various internal school logistics.

During the following four days, the course was taught to approximately twenty students ranging from 5th to 8th graders. As the topic was new, and as the course was short, we had made it an objective to have a mixed age classroom rather than a same age cohort classroom. We saw no reason to abide by the latter model, which we considered more of a remnant of the "Prussian model" (Khan, 2012) rather than a logical model.

On the fifth day, students wrote (in English and in Albanian) an open letter to the Mayor of Tirana, where they also asked for the municipality to publish their work on its website. They conducted the course evaluations, and they first organised and then attended a closing ceremony, hosted by the school, for the delivery of course certificates. The ceremony was well attended by teachers, parents, and country officials such as Albania's Deputy Minister for the Environment, Forestry, and Waters, Dr. Taulant Bino, and the media. A brief video clip of the event can be seen at: http://www.youtube.com/watch?v=H7vzZZ6zGiQ.

Figure 7.6. The first day of class.

The course was predominantly taught by one teacher who had intervening support from the other two teachers when their areas of expertise matched the topic of discussion. The teaching style was geared towards an informal round circle discussion rather than towards a classical top down model. The course was hands-on and the teachers floated around the classroom providing individual help to students, acting as half-peers with them and working in combination with one another. The class day lasted eight hours, with six hours spent exclusively in teaching and learning.

While we constantly communicated with the Albanian teachers and the school Director during the week, we requested at the outset that they would not be present in the classroom. It was our view, based on our understanding of Albanian cultural practices, that students would be more spontaneous, and that they would enjoy a higher degree of self-expression if they were not in the presence of their teachers. Our subsequent observations, during the following days, of the students' behaviour when in the presence of their teachers in hallways, re-affirmed this view.

The course was taught in English. The children did not exhibit any difficulty with English. The pace of their absorption of new concepts in

GIS or technological functions surpassed by many orders of magnitude the highest of our preconceptions, and expectations.

Coordination and Logistics Planning in Albania

It is next to impossible to describe the numerous details, and infinite strategies that we needed to employ and which frequently had to be drastically realigned overnight, or that repeatedly tested our patience, indulgence, and determination with cross-continental institutional communications, understandings, and misunderstandings with formal and informal commitments and expectations during the year in which we engaged in this project. But what is important to emphasise, and highly significant to highlight, is that more than half of our efforts in the development and the implementation of this project were spent in these activities.

Partnerships

In order to expand the resource boundaries of our course, and to connect our efforts to larger and sustained ones from other established organisations and programmes, during this phase we were successful in establishing fruitful partnerships with several not-for-profit research and educational organisations. These provided significant contributions and resources to the project. They included:

- The GISCorps, which was established in 2003, and which provides worldwide volunteer GIS services to less advantaged communities. GISCorps provided a volunteer teacher in Tirana (Dr. Jennifer Rechel from the United States Department of Agriculture in Riverside, California);
- The Green Mapping System, which was established in the mid-1990s, and which engages worldwide communities in mapping green living, nature, and cultural resources, provided us with ideas, its web resources, cartographic methods and standardised symbols, a specially made video for our students, and Green Mapping certificates. After the course, they also published the children's work products in their website;
- The Mediterranean Association for the Protection of the Sea Turtles (MEDASSET), which was established more than two decades ago, also provided a video narrative, specially made for us, about the protection of the sea turtles, brochures and educational materials that they translated expressly for us into Albanian, and

with other web resources and data that they made available to the AYFEED Project;

- The My Community, Our Earth, an international programme of the Association of American Geographers, published the children's work products on their website, supported them with additional special accreditation certificates, and also offered their political support to the project.

Results and Discussion

Cognisant that results of similar projects can be defined and described in many ways, reporting is limited to two aspects: (a) students' evaluation of the course; and (b) elements of social production employed in this project.

The research group at the University of Florida believes that this prototype course was a big success among students, parents, and school officials. It was clear to us from day one (and especially after monitoring the web traffic to the course materials that night) that these children were genuinely interested in, and highly attracted to, this course. During the closing ceremony of the course, parents told us stories of their chatting non-stop at home about what they had learned during the day and of their fervent interest in having other similar opportunities.

We started this project from a distance, and engaged in it with much trepidation. Since its onset, we were aware of many unanticipated obstacles that would have to be overcome along the way. But during the teaching week, from the beginning till the end, the children repeatedly adjusted our pre-conceptions, and exceeded our expectations in all aspects.

Below are the results from two forms of qualitative evaluations that were carried out. A structured and anonymous questionnaire was conducted at the end of the course, and a free-writing anonymous evaluation at mid-point of the course. The questionnaire asked six questions. To the first three questions, "Did you enjoy this course?", and "Would you take another similar course in the future?" and "Do you think that you learned interesting subjects during this week?", the answers were a hundred per cent "Yes." To the fourth question, "From which of the subjects did you learn the most (Environment, GIS, English or all three)?" twenty-five per cent answered GIS, while seventy-five per cent answered "all three." The other two questions sought students' opinion about the most difficult or easiest parts of the course. Eighty per cent of the answers to these two questions emphasised the ease and fascination with the course.

But it was the students' free-writing evaluations that more meaningfully captured their attitude towards the course. We asked them to write negative or positive statements about their experiences. Some answered with just a few sentences, one or two answered with full page of thoughts, and the majority wrote an average of a half page. The transcript of one evaluation follows, which best symbolises the atmosphere that we experienced.

> "I like this course very much! Why so? Because except the fact that I'm not staying home all day, I had fun, I met new people, I improved my English, I learned about something I never heard about. Also I'm learning so much beautiful things & news about animals and how can we help them. For me this course was the most beautiful thing that happened in this summer so far! Thank you!"

A collage of some of these free-writing evaluations is shown in Figure 7.7.

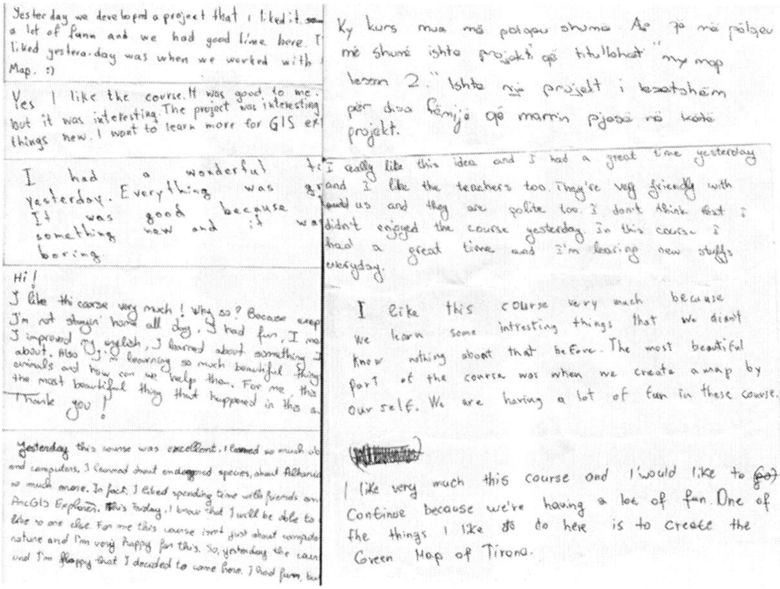

Figure 7.7. Collage from samples of free-writing student evaluations.

One particular feature about the development process of this project requires attention. And that is the manner of its production. Its production embodied many elements of what Benkler has coined contemporary "social production" as opposed to "industrial production", and which he considers "a critical long term shift caused by use of the Internet (Benkler, 2005)". The participatory nature of the production of this project, which brought together resources from several international organisations with different missions, and in different locations, exhibited many of the elements of the contemporary network cultural practices and models of creation. Each of the partner organizations made a relatively small and informal contribution to this project, but the collective sum of their input made a significant impact upon the project.

This "manufacturing process", which is "part collaborative creativity" and "part organisational style", embodied many of the elements in Shirky's definition of "the Open Source pattern" and aimed to develop more of a shared product, rather than a single ownership product. In our case, these cultural practices were blended with the traditional institutional ones, successfully blurring the boundary between the two. Indeed the informal, collaborative, not hierarchical, decentralised, and distributed contributions provided by the partner international organizations that took part in this project, were frequently more creative and efficient than parts of the institutions that were formally responsible for the development and the implementation of it. This phenomenon also overlaps with Uricchio's definition of "cultural citizenship in the age of P2P collaborative communities" as an example that supports his argument that distinguishes between a political, economic, or cultural citizenship from that of a fluid and temporary citizenship that is based on the identity of a particular community with a shared goal, and which does not necessarily "occur within the confines of the territorial nation-state" (Uricchio, 2004).

In the development and teaching of this course we sought to provide education with a focus on fundamental principles of geospatial concepts rather than just GIS computer training. We also tried to stress connections and integration of concepts across disciplines based on the experiences of daily life. We considered it important to also allocate room in the curriculum for authentic creativity (perhaps at the expense of introducing more concepts), and for student political empowerment and self-expression. Our goal was to stay clear from embracing technology for its own sake and to approach the introduction of new technology as a means for improving, and for humanizing conceptual understandings of the surrounding world.

This project was conceived as a prototype case study towards the development of a broader, short-term international curriculum that would provide integrated teaching of spatial literacy and urban sustainability subjects to young students. Work towards this effort is already underway at the University of Florida. We have formed an international research group with interest in this topic, composed of representatives from Urban and geography departments at the University of Georgia in the United States, the National University of Palestine (An-Najah), the National University of Rwanda, the National University of Uganda (Makerere), the EIS-Africa (a GIS research and education foundation with a pan-African mandate), and the University Institute of Technology in Yaracuy, Venezuela. And in Albania, it is our wish that the next step of this project will be its implementation in many other schools across the country.

References

Batty, P. 2012. *Future trends in geospatial information management: The five to ten year vision*. New York, NY: United Nations Programme on Global Geospatial Information Management (CGIM).

Benkler, Y. 2005. *The wealth of networks: How social production transforms markets and freedom*. New Haven and London: Yale University Press.

Bino, Xh. 1999. *Souvenirs et documents sur le Lycée français de Korçë: En hommage aux martirs et aux professeurs du lycee*. http://www.docstoc.com/docs/113023579/Xhuvi-BINO. (accessed June 17, 2013).

Campanile, E., Comrie, B., & Watkins, C. 2005. *Introduzione alla lingua e alla cultura degli Indo-European*. Bologna, Italy: Il Mulino.

Cook, B. A. 2001. *Europe since 1945: An Encyclopedia*. New York, NY: Garland Publishing Inc.

Gogaj, I. 2004. *Mirash Ivanaj, Personalitet i Shquar i Universit Shqiptar*. Tiranë, Albania: Erik botime.

Herrera, O., Mejías P., Gutiérrez C., & Matamoro, R. 2012. *Students as Producers and Consumers of Primary Contents Using Web 2.0 Tools*. Conference Proceedings of the 6th International Multi-Conference: Systemic, Cybernetics and Informatics. Orlando, Florida, USA.

Hyseni, S. 2012. *Histori e shkurtër e arsimit shqip*. Engjujt shiptarë. http://www.engjujtshqiptare.com/t27931-histori-e-shkurter-e-arsimit-shqip. (accessed June 14, 2013).

Ivanaj, D. 2008. *Message from the founder*. The Martin and Mirash Ivanaj Foundation. New York, NY. http://www.ivanaj-foundations.org/. (accessed June 13, 2013).

Khan, S. 2012. *The one world schoolhouse*: Education reimagined. New York, NY: Grand Central Publishing.

The National Environmental Agency of Albania (NEA). 1999. *National Report on Biodiversity Strategy and Action Plan*. Tiranë, Albania: The Institute of Biological Research and Museum of Natural Sciences.

Shirk, C. 2005. Epilogue: Open Source outside the Domain of Software. In *Perspectives on Free and Open Source Software*, eds. Feller, J., Fitzgerald, B., Hassam, S. A., & Lakhani, K. R., 483-489. Cambridge, MA: MIT Press.

United Nations Development Programme. 2011. *Human development report*. New York, NY: Colorcraft of Virginia Inc.

Uricchio, W. 2004. Cultural Citizenship in the Age of P2P Networks. In *Media Cultures in a Changing Europe*, eds. Bondebjerg, I., & Golding, P., 139-164. Bristol, UK: Intellect Press, Ltd.

White, M., Haxhiu, I., Saçdanaku, E., Petri, L., Rumano, M., Osmani, F., Vrenozi, B., Robinson, P., Kouris, S., & Venizelos, L. 2008. *Monitoring Stavnike Fish-Traps and Sea Turtle Bycatch at Patoku, Albania*. Conference Proceedings of the International Conference on Biological and Environmental Sciences. Tirana, Albania.

The World Bank. 2011. *World Development Indicators (WDI)*. Washington, DC: The World Bank Group. http://data.worldbank.org/indicator. (accessed June 14, 2013).

Zoi, N. 2000. *Një faqe historie*. Korçë, Albania: Shypshkronja Kotti.

CHAPTER EIGHT

DEALING WITH GIS IN GEOGRAPHY CURRICULA: COMPARING PORTUGAL AND TURKEY

EYÜP ARTVINLI AND CRISTIANA MARTINHA

GIS in curriculum

From ecological restoration to congressional redistricting, from military strategies to transportation planning, from emergency service deployment to the modelling of the impacts of climate change, GIS has become an indispensable tool for a wide array of practitioners and analysts in the public and private sectors (Murphy, 2007). It is now a standard item in planners' tool kits (Drummond, FAICP, 2008). The use of GIS in education also employs constructivist pedagogies, such as learning through inquiry and problem-based learning to facilitate greater engagement (Donert, 2006a, 2006b; Rød, Larsen, Nilsen, 2010; Bednarz, 2007; Milson & Earle, 2007; Madsen & Rump, 2012; Baker, 2005; Alibrandi, 2003). Hence, the educational side of GIS is very important in order to educate future generations. GIS is also starting to be used by geography teachers. However its presence in the curricula is not yet fully effective in many countries despite research confirming that its presence in the curriculum was important for its classroom application.

Lam, Lai and Wong (2009) interviewed geography teachers about their views on the inclusion of GIS in the secondary geography curriculum in Hong Kong and they concluded that the implementation of GIS in high schools depends on: i) teachers' sense of preparedness to implement GIS; ii) the perceived practicality of its use in teaching; and iii) whether GIS use was mandated in the curriculum. Goldstein and Alibrandi (2013) concluded that the inclusion of GIS in middle school curriculum had a significant effect on student achievement on final course grades in science and social studies in Florida. In Norway, Rød, Larsen and Nilsen (2010) explain how the curriculum reform of 2006 that applies to the curriculum for upper secondary schools (pupils aged 16-19 years), explicitly states that pupils should know how to use digital maps and Geographical Information Systems (GIS). They defend this in practical terms by suggesting geography teachers should start with web-based GIS applications and/or free GIS-data viewers. In Germany, Schubert and Uphues (2009) present an interesting model to serve as a general guideline for the development and implementation of local GI curricula.

The digital-earth.eu network (Comenius project) also has been doing important work on the issue of GIS in the curriculum. In fact, the position of geo-media and geoinformation in the curriculum was one of four special interest groups to explore educational perspectives. The group produced a report about curriculum issues in Europe (Donert, Parkinson and Lindner-Fally, 2012) confirming that "Geo-media and geoinformation in education are not mentioned in European policy documents, nor do they appear in many European national curriculum documents reviewed for this report" (Donert, Parkinson and Lindner-Fally, 2012: 3). Related to this perspective a guidance leaflet was produced to advise decision makers about the importance of geo-media and GIS in the curriculum Digital-Earth Network, 2012) available online at http://213.235.245.69/fileadmin/deeu_documents/D5.1_SIG4-curriculum-report-v3.pdf.

In "International Perspectives on Teaching and Learning with GIS in Secondary Schools", Milson, Demirci and Kerski (2012) invited authors to present the state of the art in their own countries. But this is an overview and discussion about the introduction and presence of GIS issues in the secondary education curriculum is lacking: this chapter aims to help to fill this gap by comparing two countries.

A comparison of GIS in the high school curricula of Portugal and Turkey

The Portuguese situation

In this analysis we refer only to secondary education. In Portugal, secondary education relates to pupils aged between 15 and 18 (10th, 11th and 12th grades). As mentioned earlier there is no overview of the way that GIS appears in the geography curriculum in Portuguese secondary education.

Geography in Portuguese secondary education is based on three different curricular subjects depending on the type of course that the pupils are taking. They are:

- Geography A – the subject that most pupils take in 10th and 11th grades. It is inserted in a scientific-humanistic course that is orientated towards following the university studies. The students take a national exam at the end of the subject;
- Geography B – is only made available to a limited number of pupils. It is based on technological courses that are not specifically orientated towards university studies: however pupils can enter university if they want and if they pass some specific national tests;
- Geography C – it is the subject that most pupils follow in the 12th grade. It is based on a scientific-humanistic course orientated to taking university studies, with a national exam taken at the end to qualify;

In Figure 8.1 we present the GIS components in the different programmes of geography in secondary school education in Portugal.

	Programs of Geography in secondary education		
	Geography A	Geography B	Geography C
Objectives and Aims	=	X	=
Theoretical Content	X	X	X
Didactical Strategies	☑	X	☑
Proposed Resources	X	=	X
Evaluation	X	X	X

Key:
☑ Presence of GIS
= Presence of related issues to GIS
X No reference to GIS

Figure 8.1. GIS in Geography Programs of Secondary Education in Portugal.

The aims and objectives of the Geography A programme indicates several "general objectives/competences" where the use of new technologies is mentioned. For instance, "To analyse the contribution of Information and Communication Technologies as a factor of development in the comprehension of and individual and social utilisation of the geographic space", "To use Information and Communication Technologies, namely informatics, telematics and multimedia", and "To connect transformations in the organisation of geographic space with the potential and limitations afforded by new Information Technologies".

In its theoretical content, there is no reference to GIS but in the 11th grade it is compulsory for pupils to develop a "case study" in their geographical work. Here we must underline the national project "Nós Propomos!" coordinated by University of Lisbon, as an example where pupils of several schools develop a case study about their home town in order to find information and present the results, where possible using GIS technology (Claudino, Martinha and Silva, 2012).

In terms of didactical strategies, the curriculum states that "Geography teaching can widely benefit from new technologies like: the access to information using, for example, data bases and geographic information systems; on its treatment, independently of the processes involved; in communications using, for example, email; in the creation and storage of digital information". Concerning the resources to be used, the curriculum only advises the use of "specific specialised software" but does not indicate the name or types of such software. The evaluation also makes no reference to GIS.

In the Geography B programme, there is no reference at all to GIS or geoinformation. The only mention occurs in the resources section where the use of the "Environment Atlas of Portugal" is proposed, as well as subject-specific software and the website of APROFGEO (Association of Geography Teachers of Portugal) where some information about this issue can be found on the Geored platform.

In the aims and objectives of Geography C, the development of transversal competences promoted by ICT is mentioned. It also presents two "general objectives/competences": i) to analyse the contribution of Information and Communication Technologies as a factor of development in the collective and individual and social uses of geographic space"; and ii) to use Information and Communication Technologies namely informatics, telematics and multimedia.

For didactical strategies, Geography C indicates that "Geography teaching can widely benefit from the use of new technologies such as: access to information using, for example, databases and geographic information systems; the independent processing and treatment of information; in communication by using, for example, email; in the creation and storage of digital information".

In terms of the resources to be used, the curriculum only recommends the use of "specialised software" and so does not indicate specific names to teachers. The evaluation processes also makes no reference to GIS or geoinformation.

We must refer also that a specific technological programme of "Geographic Information Systems" for high schools was created for the 12th grade pupils of technological course of spatial planning and environment. This subject has the following purposes: a) to become familiar with Geographic Information Techniques; b) to recognise the importance of GIS as a tool for decision support in issues about

Environment and Spatial Planning; and c) to use GIS in concrete situations dealing with Environment and Spatial Planning issues". Its main objectives are: i) to apply techniques of GIS data acquisition; ii) to use the main spatial analysis functions of GIS; iii) to produce concrete results using GIS; and iv) to connect with real situations of GIS use.

The competences to be developed through this curriculum are specifically defined as:

- Understanding the concept of geo-referenced information;
- Mastering the specific terminology of GIS;
- Distinguishing the nature of geographic information represented by matrix structures (raster and vector);
- Understanding the main functions of GIS in terms of collection, storage, management, interrogation (query), analysis and presentation of georeferenced information;
- Knowing about GIS software-based matrix systems (raster) and vector information;
- Knowing the main suppliers of cartographic information and other georeferenced information in Portugal;
- Using techniques for the acquisition of primary and secondary georeferenced information;
- Knowing the different structures of databases;
- Designing a georeferenced database;
- Performing spatial analysis procedures with a GIS, such as: overlapping layers of the different information; creating queries in the system applying restrictions; performing measurements, transformations and optimisations;
- Producing thematic maps;
- Meeting applications of Geographic Information Technologies in the areas of Environment and Spatial Planning;
- Recognising the importance of GIS to solve georeferenced problems and particularly as a tool to support decision-making on matters relating to the Environment and Spatial Planning;
- Understanding issues related to the quality of geographic information;
- Understanding the legal issues involved in the use and access to georeferenced information;
- Using Geographic Information Technologies to analyse issues in Environment and Spatial Planning at various scales;
- Performing spatial analysis enabling, for example: the observation and description of the spatial distribution of variables; quantifying areas and the population corresponding to different attributes;

simulating the diffusion processes and observing the possible affected areas; assessing the possible locations of activities; and using the constraints that may limit possible locations;
- Produce and interpret relevant maps to the Spatial Planning and Environment.

In terms of resources the programme presents many references to specific tools and technologies like: Geomedia Professional Student License, Mapinfo Proviewer, ArcExplorer, ArcReader, INOVAGIS, SPRING, The Environmental Systems Research Institute (ESRI), Intergraph, Microsoft and Sun. It also presents several information sources from Portugal and relevant websites, which can be used to explore information and data from Portugal and other countries.

In this technologically-orientated spatial planning and environment course the subject of "Techniques of Spatial Planning" must be closely referred to. Reference to the use of GIS software is found in one of its themes, "Landscape Representation".

It is important to underline that these subjects need access to relevant tools for students of secondary education to work as geographers and citizens and they should allow to students and teachers to work in order to implement the geography benchmarks presented by HERODOT network-http://www.herodot.net/Geography-benchmark.html.

GIS in curriculum of high school education in Turkey

Two hours per week geography teaching in Turkish high schools is compulsory in grade 9 and 10 for every kind of high school. In 11th and 12th grades, while geography is an optional subject for some types of school like a science high school, it is compulsory for other types like a social studies high school. It is also optional in the department of science and maths in normal (or Anatolian) high schools but it is compulsory in the department of social studies of normal (or Anatolian) high schools.

In 2005 a new geography curriculum was introduced in secondary schools in Turkey. Before this curriculum, GIS was only discussed at a very rudimentary level in some textbooks (Demirci and Karaburun, 2009). Many Turkish researchers focused on the importance of GIS use in schools in Turkey (Demirci, 2008, 2009, 2012; Ozgen, 2009; Artvinli, 2010 and Incekara, 2012). Subsequently, the Turkish geography curriculum was again revised in 2010 and 2011. The 2011 geography curriculum indicated that "this geography curriculum supports the use of ICT during the teaching of geography lessons". As a result, the use of Geographical Information System (GIS) applications is advised during the teaching of some standards. Teachers can develop or investigate some examples of GIS lessons with qualified technical software and hardware in schools for GIS (Article 18, CDOP; MoNE, 2011: 6)".

Within the 2005 curriculum, GIS was recommended 20 times as a tool for activity development (Demirci, 2008). But as can be seen in the tables below, after renovation of the geography curriculum in 2011, GIS usage is recommended 27 times within the whole curriculum. This change in the curriculum drew the attention of geography teachers to GIS and motivated them to learn more about the possibilities of GIS and its use in their lessons. Moreover, every geography standard was associated with geographic skills, which are described inside the geography curricula. GIS usage can be identified for each standard in the curriculum. These standards can easily be compared with geographic skills to be attained. In this way we can understand which geographic skills can be most developed by using GIS in the curriculum. This allows the analysis of standard-geographic skill balance within GIS concepts.

Number of Standard	Learning Area of the Standard	Content of Standard	Given Related Geographic Skills For This Standard In Curriculum
A.9.4.	Natural Systems	Students compare the techniques and methods to transfer information on to maps according to "usage" aims.	1. Map reading 2. The effective usage of primary and secondary data sources, (never described to teachers inside of curriculum) 3. Using evidence
A.9.5.	Natural Systems	Students make inference about the properties of "time" and "space" by using the tools which compose the frame of reference and maps.	1. Map reading 2. Inquiry
A.9.6.	Natural Systems	Students recognize main landforms on a map by drawing using contour lines.	1. Map reading 2. Fieldwork 3. Observation

Figure 8.2. 9th grade Geography curriculum standards where GIS usage is advised.

As can be seen in Figure 8.2., GIS usage is advised only within the learning area of natural systems in the 9th grade geography curriculum. It is important to introduce GIS to students in the first year of geography in high schools. In this way students start to be familiar with GIS and its usage for better geographical analysis.

Schubert & Uphues (2009) proposed four levels of using GIS in schools (Figure 8.3.). Four levels are separated for using GIS according to this pyramid. Two of them are basic and teacher centred on year 6/7 students (12-13–year-old). The third level is for year 8/9, and the top level is for year 10/11 (15-16 years). When we analyse the standards of 9th grade geography curricula where GIS usage is advised, we can see the result in Figure 8.3.

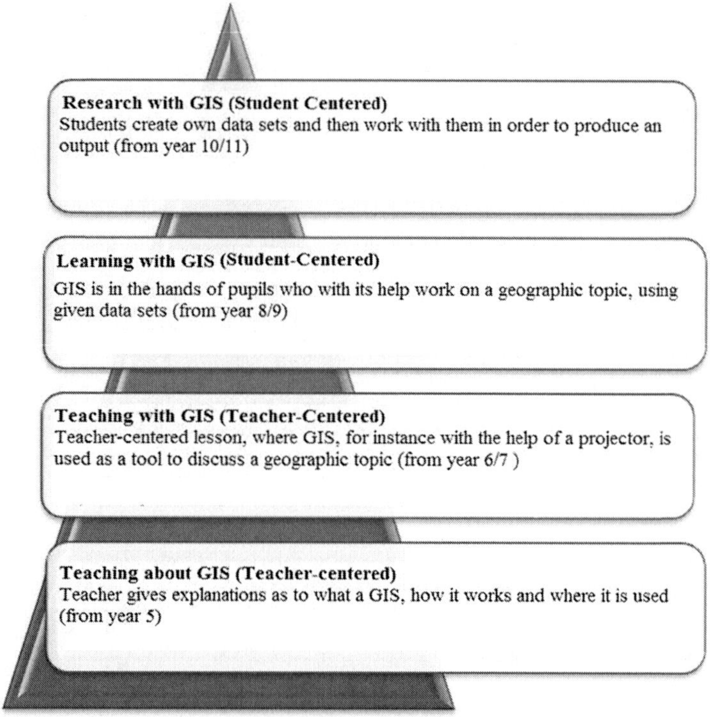

Figure 8.3. The level of GIS use in schools.
(Schubert & Uphues, 2009)

Standard Number	Quality of proposed GIS Using	Approach
A.9.4.	Teaching about GIS	Teacher-centred
A.9.5.	Teaching with GIS	Teacher-centred
A.9.6.	Teaching with GIS	Teacher-centred

Figure 8.4. Quality of GIS Using for Standards of 9th grade Geography curricula.

In Figure 8.4. we present the quality of GIS use for 9th grade according to Figure 8.3. The quality of the proposed GIS usages in standards is teacher-centred as per Figure 8.4. But according to the pyramid in Figure 1 the "teacher-centred" level is for the year 5 and year 6/7 (up to 13 years old). In this respect it can be suggested that the 9th grade geography curriculum should include some complicated GIS use based on "learning GIS" to investigate, analyse and conclude geographical phenomena.

On the other hand it is clear that merely introducing GIS usage in some geography lessons in Turkish high schools is not enough when we consider the fast development of GIS usage in other school subjects like mathematics, history, etc. Moreover, there is a lack of GIS usage in other learning areas of geography in this grade. This situation has the potential to make geography appear as a memorisation subject in the eyes of students as this kind of usage of GIS only presents the visualisation of physical features and does not include the use of analytical tools. There is no well-balanced development of GIS use between the learning areas of the 9th grade geography curriculum.

On the other hand, in this grade the standards where GIS usage is advised are associated with some geographical skills like map reading, using evidence, inquiry, fieldwork and observation. At this level, GIS-related standards stress developing "map reading" geographic skills as per Figure 8.2.

Number of Standard	Learning Area of the Standard	Content of Standard	Given Related Geographic Skills For This Standard In Curriculum
A.10.2	Natural Systems	Students correlate the distribution of earthquake zones, hot water sources, volcanoes with plate tectonics.	1. Map reading 2. Observation 3. Inquiry 4. Using evidence 5. Perception of change and continuity
B.10.1.	Human Systems	Students make inference about characteristics and importance of the population.	1. Inquiry 2. Perception of change and continuity 3. Interpretation of tables and graphs
B.10.2.	Human Systems	Students investigate the distribution of population in the world and its factors.	1. Map reading 2. Inquiry 3. Interpretation of tables and graphs 4. Perception of changes and continuity
B.10.3.	Human Systems	Students interpret the changes of world population from a historical perspective.	1. Inquiry 2. Perception of time 3. Perception of change and continuity 4. Using evidence
B.10.4.	Human Systems	Students conclude about population structure by making population pyramids.	1. Interpretation of tables and graphs 2. Perception of change and continuity 3. Using evidence
B.10.5.	Human Systems	Students interpret the factors on the change of population grow rate.	1. Map reading 2. Inquiry 3. Interpretation of tables and graphs 4. Perception of change and continuity
B.10.9.	Human Systems	Students analyse economic activities according to their basic features.	1. Map reading 2. Observation 3. Fieldwork 4. Making and interpreting tables, diagrams and graphs
B.10.10.	Human Systems	Students correlate development levels of countries with data of the proportional distribution of types of economic activities.	5. Map reading 6. Observation 7. Fieldwork 8. Making and interpreting tables, diagrams and graphs
C.10.4.	A Spatial Analysis: Turkey	Students conclude about the factors that effect of climate in Turkey.	1. Map reading 2. Observation 3. Inquiry 4. Making and interpreting tables, diagrams and graphs 5. Perception of change and continuity

C.10.5.	A Spatial Analysis: Turkey	Students conclude about the characteristics of climatic elements in Turkey.	1. Map reading 2. Observation 3. Inquiry 4. Making and interpreting tables, diagrams and graphs 5. Perception of change and continuity
C.10.6.	A Spatial Analysis: Turkey	Students conclude about characteristics of climate types in Turkey by using evidence of climate.	1. Map reading 2. Inquiry 3. Making and interpreting tables, diagrams and graphs 4. Perception of time 5. Perception of change and continuity 6. Using evidence
C.10.15.	A Spatial Analysis: Turkey	Students give examples about the cities of Turkey according to their characteristics.	1. Map reading 2. Inquiry 3. Using evidence
C.10.16.	A Spatial Analysis: Turkey	Students analyse the distribution and structural characteristics of population in Turkey.	1. Map reading 2. Inquiry 3. Making tables, diagrams and graphs 4. Perception of change and continuity
C.10.17.	A Spatial Analysis: Turkey	Students query in terms of social and economic factors of the population in Turkey in a historical course.	1. Map reading 2. Inquiry 3. Making and interpreting tables, diagrams and graphs 4. Perception of change and continuity
D.10.1.	Environment and Society	Students analyse natural hazards according to their reasons of appears, characteristics, contribution and effects on people.	1. Map reading 2. Observation 3. Fieldwork 4. Inquiry 5. Making and interpreting tables, diagrams and graphs

Figure 8.5. 10th grade Geography curriculum standards where GIS usage is advised.

In Figure 8.4. we can observe a better balanced distribution of potential GIS usage when compared with grade 9 learning areas. In this case, most GIS usage is advised for "Human Systems" learning areas and "A Spatial Analysis: Turkey". At this level of the curriculum "Physical Systems", "Global Environment: Regions and Countries" and "Environment and Society" are not considered suitable for GIS teaching and learning. However this doesn't mean that GIS usage is included in geography lessons. This is the only curriculum suggested for teachers to use GIS as a teaching learning tool for students for determined standards. On the other hand, most research-related GIS usage and applications in classrooms argue that GIS is used less during teaching and learning processes for various reasons including teacher training, non-user friendly software, and lack of time to conduct GIS applications during lessons (Demirci, 2008, Ozgen, 2009; Artvinli, 2008, 2010 and Incekara, 2012). On the other hand, the 10th grade curriculum has the most standards, which advise potential usage of GIS during teaching and the learning process for students. But there is still a lack of advising GIS usage in some important learning areas like regions and countries. It should be mentioned that this requires high-thinking and critical skills for students to be able to make analyses, but most standards to apply potential GIS are not based on higher critical thinking skills.

Standard Number	Quality of proposed GIS Using	Approach
A.10.2	Teaching with GIS	teacher-centred
B.10.1.	Teaching about GIS	teacher-centred
B.10.2.	Teaching about GIS	teacher-centred
B.10.3.	Learning with GIS	student-centred
B.10.4.	Research with GIS	student-centred
B.10.5.	Research with GIS	student-centred
B.10.9.	Learning with GIS	student-centred
B.10.10.	Learning with GIS	student-centred
C.10.4.	Learning with GIS	student-centred
C.10.5.	Learning with GIS	student-centred
C.10.6.	Teaching with GIS	teacher-centred
C.10.15.	Teaching with GIS	teacher-centred
C.10.16.	Research with GIS	student-centred
C.10.17.	Learning with GIS	student-centred
D.10.1.	Teaching about GIS	teacher-centred

Figure 8.6. Quality of GIS Using for Standards of 10th grade Geography curricula.

In Figure 8.6. we present the quality of GIS usage for the 10th grade, according to Figure 1. The quality of the proposed GIS usages in standards is student-centred (6 standards) and teacher-centred (6 standards). According to these results, most of the 10th grade curriculum has a balance between student-centred and teacher centred potential GIS using activities. But when we consider Figure 8.3. and its levels for GIS use it should include more standards for "research with GIS and "learning with GIS" activities in this grade.

When we check the suggested geographic skills, to develop GIS-related standards in the 10th grade curriculum it can be seen that "map reading" is still the most important geographical skill. On the other hand, "inquiry" and "making and interpreting tables, diagrams and graphs" are the second most important skills. It can be seen that nine standards have the potential to apply student-centred GIS activities in this grade, and it is useful to advise inquiry skills be developed within GIS activities.

Number of Standard	Learning Area of the Standard	Content of Standard	Given Related Geographic Skills For This Standard In Curriculum
B.11.2.	Human Systems	Students interpret global and regional effects of cities by exemplifying them according to their functional features.	1. Map reading, 2. Perception of change and continuity, 3. Using evidence
B.11.4.	Human Systems	Students analyse the effects of natural elements on production, distribution and consumption.	1. Map reading 2. Fieldwork 3. Inquiry 4. Making and interpreting tables, diagrams and graphs 5. Using evidence
B.11.5.	Human Systems	Students analyse the effects of human elements on production, distribution and consumption.	1. Map reading 2. Fieldwork 3. Inquiry 4. Making and interpreting tables, diagrams and graphs 5. Using evidence
B.11.6.	Human Systems	Students analyse of connection between the sectors of distribution and consumption in terms of effect of economy.	1. Map reading 2. Fieldwork 3. Inquiry 4. Making and interpreting tables, diagrams and graphs 5. Using evidence
C.11.2	Global Environment: Regions and Countries	Students relate areas of markets, production, raw material and trade between countries and regions.	1. Map reading 2. Inquiry 3. Perception of change and continuity 4. Making and interpreting tables, diagrams and graphs

Figure 8.7. 11th grade Geography curriculum standards where GIS usage is advised.

Figure 8.7. shows how often 11th grade students are advised to learn with GIS geography lessons and with what subjects/standards. While only physical systems was advised for using GIS in the 9th grade geography curriculum, in grade 11 only "Human Systems" are advocated. On the other hand, the level of standards is higher for students as most ask for "analysis".

Standard Number	Quality of proposed GIS Using	Approach
B.11.2.	Learning with GIS	student-centred
B.11.4.	Research with GIS	student-centred
B.11.5.	Research with GIS	student-centred
B.11.6.	Research with GIS	student-centred
C.11.2	Research with GIS	student-centred

Figure 8.8. Quality of GIS Using for Standards of 11th grade Geography curricula.

In Figure 8.8. we present the quality of GIS usage for 11th grade according to Figure 8.3. The quality of the proposed GIS usages in standards is student-centred. According to Figure 8.8. the 11th grade curriculum has a student-centred potential GIS using activities.

When we check the geographic skills suggested for developing GIS-related standards in the 11th grade curriculum, it can be seen that "map reading" is still the most important geographical skill in this level too. "Inquiry" and "making and interpreting tables, diagrams and graphs" skills are the second most. All of the quality of activities and standards are based on student-centred approach.

Number of Standard	Learning Area of the Standard	Content of Standard	Given Related Geographic Skills For This Standard In Curriculum
C.12.4.	A Spatial Analysis: Turkey	Students analyse trade centres, the goods which are used in trading and their flow directions in terms of changes and continuity in Turkey.	1. Map reading 2. Fieldwork 3. Inquiry 4. Making and interpreting tables, diagrams and graphs 5. Perception of change and continuity
C.12.5.	A Spatial Analysis: Turkey	Students analyse external trade and the position of Turkey in world markets in terms of the goods which are used in trade.	1. Map reading 2. Fieldwork 3. Inquiry 4. Making and interpreting tables, diagrams and graphs 5. Perception of change and continuity
C.12.11.	A Spatial Analysis: Turkey	Students create scenarios based on population projections of Turkey.	1. Observation 2. Inquiry 3. Time perception 4. Perception of change and continuity 5. Using evidence
C.12.5.	A Spatial Analysis: Turkey	Students evaluate global and regional effects of location for a country.	1. Map reading, 2. Observation, 3. Inquiry, 4. Perception of change and continuity, 5. Using evidence

Figure 8.9. 12th grade Geography curriculum standards where GIS usage is advised.

In the final grade of high school, before students start university, the complexity of advised potential of GIS usage is at a relatively advanced level. The first two ask students to "analyse" information with GIS. The third one asks them to "create" scenarios and the last one asks them to use GIS to "evaluate" global and regional effects of location for a country. On the other hand, only 4 of the 37 standards include GIS teaching and learning opportunities. Normally, GIS could be used to support effective learning and teaching in all kinds of geography standards. There is thus no reason to only connect the use of GIS to specific standards. The general aims of a modern geography curriculum should include encouraging students and teachers to use GIS in all phases of teaching and learning.

Standard Number	Quality of proposed GIS Using	Approach
C.12.4.	Research with GIS	student-centred
C.12.5.	Research with GIS	student-centred
C.12.11.	Research with GIS	student-centred
C.12.5.	Research with GIS	student-centred

Figure 8.10. Quality of GIS Using for Standards of 12th grade Geography curricula.

In Figure 8.10. we present the quality of GIS usage for the 12th grade, according to Figure 8.3. The quality of the proposed GIS usages in standards is student-centred as per Table 9. According to Table 9, the 12th grade curriculum has a student-centred potential GIS using activities.

When we check the geographic skills suggested for developing GIS-related standards in the 12th grade curriculum, it can be seen that "inquiry" and "perception of change and continuity" are the most important geographical skills in this level. "Making and interpreting tables, diagrams and graphs" and "making and interpreting tables, diagrams and graphs", "observation" and "using evidence" skills are other important skills to develop for these standards. All of the quality of activities and standards are based on a student-centred approach in the 12th grade curriculum. On the other hand, only four standards out of 37 are advised for GIS usage for teaching them to students, but the first two ask them "analyse" with GIS to students. The third one requires "creating" scenarios and last one requires using GIS to "evaluate" the global and regional effects of location belonging to a country.

As a final comparison we present a general view of GIS education in Turkish geography curricula.

	Programs of Geography in high school education			
	Geography 9	Geography 10	Geography 11	Geography 12
Aims and objectives	X	X	X	X
Theoretical Content	☑	☑	☑	☑
Didactical Strategies	☑	☑	☑	☑
Proposed Resources	X	X	X	X
Evaluation	X	X	X	X

Key:
☑ Presence of GIS
X No reference to GIS

Figure 8.11. GIS in Geography Programmes of High School Education in Turkey.

Geography curricula have some common and separated parts for every grade. Most of its content is common to every level of teaching geography, like aims, explanation of curriculum, vision and principles, approach to curriculum, main components (skills, concepts, values and attitudes), learning areas, measurement and assessment (all of these topics are explained in the first 77 pages of the curriculum, and the remainder, up to page of 135 covers standards for the 9th to 12th grades). Separate parts include only standards for 9th to 12th grades. It is thus not very suitable to analyse the common parts of it for each grade or level. But we can in any case conclude that the geography curricula of Turkey have these drives and challenges in Figure 8.11.

The geography curricula have 14 main goals but none focus on GIS or technology use in lessons. But in its theoretical content (standards) GIS use is advised 27 times for some standards shown in previous figures. It is also suggested a number of times within the curriculum that GIS be used as a teaching and learning tool where technological opportunities are suitable for it inside the classroom. On the other hand, there is not enough information about GIS and how to use it within the classroom, either in practice or as a tool. Moreover, the curriculum has 36 pages on making evaluation and assessments at every level of geography lessons, but there is no suggestion or advice on using GIS or evaluating students if they learn enough about GIS or research with GIS in their lessons.

Conclusions

If we analyse the Portuguese and Turkish situation with respect to the HERODOT network benchmark statement "GIS in Secondary School Education: a benchmark statement"—http://www.herodot.net/Geography-benchmark.html—the Portuguese curriculum would appear to be better orientated to fill its requirements than the Turkish one.

Looking at and comparing the presence of GIS in the geography curriculum of secondary education in Turkey and in Portugal we conclude that this issue is referred to more extensively in the Portuguese geography curriculum of secondary education (2001 and 2002) than in the Turkish one (2011). It seems that in Turkey GIS is not explicitly mentioned in the geography curriculum. By comparing the curriculum with the possible uses of GIS in Education (Figure 8.3.), we can conclude that in Portugal the use of GIS is student-centred, whereas in Turkey a teacher-centred approach is indicated, for example only "teaching about GIS, and teaching with GIS" is advised. We can conclude that main reason for this situation has been the absence of enough suitable teacher education and training in GIS use.

It can be argued that the Turkish geography curriculum does not widely support students who graduate from high schools to participate in public decision-making through the uses of spatial information and visualisation. For example, it is not enough to raise the awareness of students of the ways of maintaining and building their own GIS knowledge and skills, it should be clear that the uses of GIS support student understanding of the basic purpose and application of GIS to deal with interdisciplinary real world problems.

The Turkish national curriculum for geography needs to address this situation. It should be revised in the near future to introduce student-centred, active GIS use in high school education in a more effective way. In fact GIS could be used to support learning and teaching in all areas of geography. There is thus no reason to indicate GIS use only with specific national standards. The aims of the geography curriculum should encourage students and teachers to use GIS in every phase of high school geography, with a focus on higher-order spatial thinking skills.

In Portugal, on the other hand, despite the fact that GIS is very visible in the curriculum, its effective application in practice has many obstacles that need to be solved by teacher training in GIS. So, we contend that it is very relevant (and indispensable) that the curriculum in different countries should make reference to GIS by specifically explaining the way it should be used, but its implementation will be only successful if appropriate teacher training is provided and good resources are produced to be used in classrooms. But the presence of GIS in curriculum is, without doubt, the beginning of the process of GIS implementation in schools.

When considering the status of GIS in the Portuguese secondary geography curricula, the most important feature was that there were far fewer references when compared to the Turkish case. The main reason for this would appear to be the fact the geography programmes are much older, they date from 2001 and 2002 and were developed before this time. The world of geographical information and geo-technologies has significantly developed in the past ten years and these changes are not reflected in the official documents. A subject like geography, which is highly connected to the modern world, requires a more contemporary approach.

In Portugal during 2012-2013 the Ministry of Education started to make an updated list of "outcomes" for the subjects of basic education with the aim of developing a competence-based approach. It is expected that in next academic year (2013-2014) the Ministry will publish the outcomes for the secondary education subjects. It is expected that the issues concerning GIS will appear there. It is hoped that research and publications about the importance of GIS in education will be considered and that innovative projects using GIS developed in Portugal but also from the European digital-earth network can provide a strong "lobby" to "put" GIS clearly in the curricula of secondary geography secondary education in Portugal.

The most recent textbooks include many references to GIS use. Sometimes the textbooks are more important for the teachers to use than curricula programmes, as they are more up-to-date. Some information about this specific issue can be found in Martinha (2012, 2013a, 2013b). Guidance on geographical education is necessary. We therefore suggest that a new (or updated) charter for geography education is needed, as produced by the Commission of Geographical Education of the IGU, clearly stating that GIS and geo-media should be part of the tools to be used in geographical education.

References

Alibrandi, M. (2003). *GIS in the Classroom: Using Geographic Information Systems in Social Studies and Environmental Science*, Portsmouth, New Hampshire: Heinemann.

Alves, M., Brazão, M. & Martins, O. (coord.) (2001) - *Programa de Geografia A*. Lisboa, Ministério da Educação – Departamento do Ensino Secundário.

Artvinli, E. (2010). The Contribution of Geographic Information Systems (GIS) to Geography Education and Secondary School Students' Attitudes Related to GIS, *Educational Sciences: Theory & Practice*, 10 (3): 1277-1292.

—. (2009). Approaches of Geography Teachers to Geographical Information Systems (GIS), *Balıkesir University Sosyal Bilimler Enstitüsü Dergisi*, 12 (22): 40-57.

Baker, T. (2005). Internet-based GIS mapping in support of K–12 education. *The Professional Geographer* 57 (1): 44-50.

Bednarz, S. (2007). Mapping the way forward in an uncertain world: Spatial thinking and Geography, in: S. Catling & E. Taylor (Eds) *Changing Geographies: Innovative Curricula (Proceedings of the London Conference)*, pp. 13–26, London: Herodot and International Geographical Union, Commission for Geographical Education, University of London, Institute of Education.

Claudino, S., Martinha, C. & Santos, R. (2012) – "Projecto "NósPropomos! Cidadania e Inovação na Educação Geográfica": A construção de uma ativa cidadania territorial" in Royé, D. et al. - *XIII*

Coloquio Ibérico de Geografía - Respuestas de la Geografía Ibérica a la crisis actual. Santiago de Compostela, Meubook, p. 1633-1642.

Demirci, Ali (2012) – "Turkey: GIS for Teachers and the advancement of GIS in Geography Education" in Milson, A., Demirci, A., Kerski, J. (ed.) – *International Perspectives on Teaching and Learning with GIS in Secondary Schools*. New York, Springer, p. 271-281.

Demirci, A. & Karaburun, A. (2009). How to Make GIS a Common Educational Tool in Schools: Potentials and Implications of the GIS for Teachers Book for Geography Education in Turkey, *Ocean Journal of Applied Sciences* 2(2):205-215.

Demirci, A. (2008). *Öğretmenler icing CBS: Coğrafi bilgi sistemler*i. İstanbul: Faith Üniversitesi.

Digital-earth.eu network – SIG4 (2012) - *Geo-Media in the Curriculum: for a better world!*. [available in http://213.235.245.69/fileadmin/deeu_documents/SIG4_leaflet_curriculum.pdf in 19.05.13].

Donert, K. (2006a). The use of ICT eLearning in Geography: Purnell, K.; Lidstone, J. & Hodgson, S. (Eds.), *HERODOT Perspectives in European Higher Education, Changes in Geographical Education: Past, present and Future*, IGU, CGE Symposium (pp.146-154). Brisbane.

—. (2006b). Geoinformation in European education: A Revolution Waiting to Happen, *Teaching Geography in and about Europe*, http://www.herodot.net/conferences/torun2006/teaching-europe.pdf#page=117, accessed: 10.06.2008.

Donert, K., Parkinson, A. & Lindner-Fally, M. (2012) – *Curriculum Opportunities for GeoInformation in Europe* in http://213.235.245.69/fileadmin/deeu_documents/D5.1_SIG4-curriculum-report-v3.pdf

Drummond, W. & FAICP, S. P. F.(2008). The Future of GIS in Planning: Converging Technologies and Diverging Interests, *Journal of the American Planning Association*, 74:2, 161-174.

Ferreira, A. & Ferreira, F. (Ed.) (2006) – *Programa de Sistemas de Informação Geográfica – 12.º ano*. Lisboa, Ministério da Educação – Direcção-Geral de Inovação e de Desenvolvimento Curricular.

Goldstein, D. & Alibrandi, M. (2013) - "Integrating GIS in the Middle School Curriculum: Impacts on Diverse Students' Standardized Test Scores", *Journal of Geography*. 112:2, 68-74.

Incekara, S. (2012) – "Do Geographic Information Systems (GIS) Move High School Geography Education Forward in Turkey? A Teacher\'s

Perspective", in *Application of Geographic Information Systems, InTech Publication*, 01/10/2012, pp. 83-100.

Lam, C.-C., Lai, E. & Wong, J. (2009) – "Implementation of geographic information system (GIS) in secondary Geography curriculum in Hong Kong: current situations and future directions" in *International Research in Geographical and Environmental Education*. 18:1, 57-74.

Lúcio, J., Brazão, M. & Martins, O. (Ed.), (2004) – *Programa de Technical de Ordenamento do Território – 10.º e 11.º anos*. Lisboa, Ministério da Educação – Departamento de Ensino Secundário.

Madsen, L. M. & Rump, C. (2012). Considerations of How to Study Learning Processes when Students use GIS as an Instrument for Developing Spatial Thinking Skills, *Journal of Geography in Higher Education*, 36:1, 97-116.

Martinha, C. (2012) – "A abordagem dos SIG nosmanuaisescolares de Geografia - notas de umacomparaçãointernacional" in Royé, D. et al. - *XIII Coloquio Ibérico de Geografía - Respuestas de la Geografía Ibérica a la crisis actual*. Santiago de Compostela, Meubook, p. 1654-1662.

—. (2013a) - "O desenvolvimento do spatial thinking a través de manuais escolares de Geografia - notas de uma comparação internacional e implicações para as políticas em Educação Geográfica em Portugal" in Fernandes, J., Cunha, L. & Chamusca, P. (Ed.), *Geografia & Política, Políticas e Planeamento / Geography & Politics, Policies and Planning*. Porto, FLUP/CEGOT, p. 408-414.

—. (2013b) - "GIS presence in Geography textbooks - a highway to spatial thinking development?" in *Journal of Research and Didactics in Geography* (J-READING). 1, 2, June, p. 57-66.

Martins, O. Alberto, A. & Além, M. (2001) – *Programa de Geografia B – 10.º ano*. Lisboa, Ministério da Educação – Departamento do Ensino Secundário.

Martins, O. Alberto, A. & Além, M. (2001) – *Programa de Geografia B – 11.º ano*. Lisboa, Ministério da Educação – Departamento do Ensino Secundário.

Martins, O. Alberto, A. & Além, M. (2002) – *Programa de Geografia B – 12.º ano*. Lisboa, Ministério da Educação – Departamento do Ensino Secundário.

Martins, O. Alberto, A. & Alves, M. (2002) – *Programa de Geografia C*. Lisboa, Ministério da Educação – Departamento do Ensino Secundário.

Milson, A. J., ; Earle, B. .D. (2007). Internet-based GIS in an inductive learning environment: A case study of ninth grade Geography students. *Journal of Geography* 106 (6): 227–237.

Milson, A., Demirci, A. & Kerski, J. (eds.) (2012) – *International Perspectives on Teaching and Learning with GIS in Secondary Schools*. New York, Springer.

MoNE (2011). *Turkish Geography Curriculum and Teaching Guide*, Ministry of National Education, Ankara.

Murphy, A. (2007). Geography's place in higher education in the United States, *Journal of Geography in Higher Education*, 31(1): 121–141.

Ozgen, N. (2009), The Functionality of a Geography Information System (GIS) Technology in Geography Teaching: Application of a Sample Lesson, *Educational Sciences: Theory & Practice*, 9, (4): 1879-1894.

Rød, J., Larsen, W. & Nilsen, E. (2010) – "Learning Geography with GIS: Integrating GIS into upper secondary school Geography curricula" in *NorskGeografiskTidsskrift - Norwegian Journal of Geography*, vol. 64, Issue 1, p. 21-35.

Schubert, J. & Uphues, R. (2009) – "Learning with geoinformation in German schools: systematic integration with a GIS competency model" in *International Research in Geographical and Environmental Education*. vol. 18, issue 4, p. 275-286.

CHAPTER NINE

USING PARTICIPATORY PROCESSES WITH YOUNG PEOPLE FOR THE DEFINITION OF CULTURAL HERITAGE: A CASE STUDY OF GENOA

LORENA ROCCA, LIVIO CHIARULLO, PIERO MORSELETTO AND GIOVANNI DONADELLI

Introduction[1]

Yi Fu Tuan's (2008) humanistic geography approach, based on the importance of people's perceptions, creativity, and personal beliefs, as well as on the influence of experience in developing attitudes towards places, is the main source of inspiration for the project introduced here. According to this approach, both the geographical and the behavioural environment (Lacoste, 1976) are to be considered separately: such a separation of spatial and emotional aspects, as well as the fact that we have to start from people's perceptions, makes it clear that future citizens should become protagonists of the process of recognising territorial objects. Such a process overcomes the "objectively recognisable heritage" construct and creates a new representation of identity. In other words, the static rules that define accessibility in the top-down approach ("this is the monument, you have to visit it") are reversed into a bottom-up approach

---

[1] The present contribution is the result of collaboration by the authors. In particular, sections 1 and 2 were written by Lorena Rocca; 3 by Lorena Rocca in collaboration with Livio Chiarullo and Giovanni Donadelli, sections 4 and 5 by Lorena Rocca with Piero Morseletto. Our thanks go to Aline Chiabai for sharing the project, and to Cristina Minelle for the translation.

where students participate together in what truly matters in their region; in doing so, they also become genuine promoters for local development (Costa 2005; Costa 2008).

The reasons for a bottom-up approach: the importance of participation

In this context, a participatory approach with students makes it possible to include different sensibilities while maximizing the citizens' interest in culture (Armstrong 2010; Eaves 2010). Moreover, stakeholders are involved in heritage management activities (Bramwell and Lane 2000; Buhalis 2003; Buhalis and Pistidda 2008; Mitsche, Reino, Knox, and Bauernfeind 2008) both by collecting information and preferences from the city "users", and, at a more complex level, by opening a dialogue with stakeholders for joint management (Carter and Belanger, 2005). In addition, participatory approaches are particularly relevant to reduce conflicts and facilitate solution-finding in the school. Framing and re-framing problems from different perspectives helps to elaborate more suitable and equitable outcomes (Schegg et al., 2008).

In the long run, looking through the lens of participation is most helpful in disentangling problems that arise from the intensity and the articulation of interactions, such as in a region, in cultural heritage, or in an urban landscape (Carugati, Hadzilias and Demoulin, 2005). Moreover, a participatory approach developed in the context of the Strategic Management (Freeman, 1984) continues to inspire scholars and students concerned with a more practical view of how business and capitalism actually work (Freeman and McVea, 2001). Thanks to this contribution several innovations were possible in management practices: many of these innovations have fundamentally changed the relationships between the organization and its employees, customers, suppliers, and other stakeholders (Anthony, Atkinson, Waterhouse and Wells, 1997).

Tools facilitating participation: e-services to enhance the city

Information and Communication Technologies (ICT) can play a crucial role in helping students to oversee and integrate operations (Alford, 2008). Therefore, ICT solutions have been increasingly employed to improve project management and deployment while reducing costs and optimizing the use of resources. E-services, identified as a set of actions mediated by information technology, proved to be most suitable to be employed in activities of territorial governance and interaction among a plurality of actors (Asgarkhani 2005; Ciborra 2002).

Some studies on online forms of participation (OECD 2003; Vedel 2003; Macintosh 2004) observe a development following three main directions:

1) At the informative level: informing about issues such as sightseeing, transport and cultural activities; to this effect, not only is the Internet a suitable tool to increase the quantity of the information provided, but it also improves its quality (Hanzl 2007). Through the Web it is possible to circulate copies of original documents, texts which—because of their size—are hardly read, summaries of official documents, and so on, but we also have links referring to specialised pages, to different sites dealing with the same piece of news, to glossaries helping to understand the text; all this with extremely reduced costs and in quite a short time.

2) At the communicative level: publishing reports announcing results, and giving the opportunity to users to upload in the system their own contribution (photos or images, for example); the Internet is seen as a "place" where it is possible to open dialogues and confrontations among citizens and, among the latter and the public decision-makers (González, Gilmer, Foley, Sweeney and Fry, 2008).

3) At the participatory level: involving citizens and stakeholders actively in the territorial management of activities (Kalay, Kvan and Affleck, 2007).

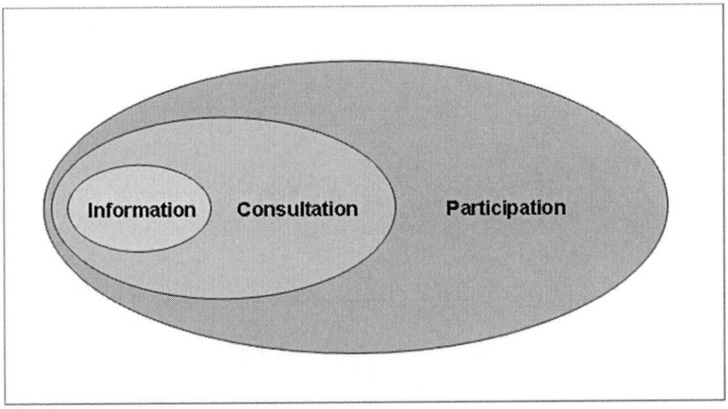

Figure 9.1. Three levels of participation (EC, 2003).

These three levels (Figure 9.1.) levels often occur at the same time and co-produce results so that—in a virtuous circle—the outcome of a level enforces or contributes developing another.

Possible effects of participative processes

Participative, bottom-up processes with students enhances cultural heritage and allows them to foresee two levels of change: on one hand, future citizens can be given the opportunity to discuss their ideas of the city and they as a result they can become aware of its riches and resources; on the other hand, the resident students can outline alternative tourist itineraries, maybe less popular but full of meaning for those who live there. Furthermore, the shared bottom-up contributions, supported by the local authorities, could give a new boost to the local development through the establishment of an alternative tourism. If, besides mapping the sites and identifying the most meaningful places (where), we manage to find the e-services and, as a consequence, suitable web tools to increase the accessibility of the places (how), we can succeed in creating specific tools to develop tourism potential.

Given the importance of information, communication and participation through the Web, the hypothesis of this study was that an increased sensitisation of the local authorities, and the creation of appropriate tools facilitating on-line participation, could also increase the opportunities of dialogue, exchange and involvement. As regards cultural heritage identification, a bottom-up approach with students, which implies a dialogue made up of several voices, allows them to identify a higher number of sites and some specific e-services useful to improve their accessibility.

The case study[2]

This chapter analyses the initiative developed in the city of Genoa as part of the ISAAC project (Integrated e-Services for Advanced Access to Heritage in Cultural Tourist Destinations), a three-year research initiative to promote a novel, stakeholder-relevant, technological environment for

[2] The study is part of the European 'Framework Sixth' Project ISAAC: "Integrated e-Services for Advanced Access to Heritage in Cultural Tourist Destinations", a multi-disciplinary research initiative aiming to enhance cultural assets as tourism resources through user-friendly and stakeholder-relevant integrated e-services in urban tourist destinations (Chiabai, 2008).

cultural heritage content. The project, financed by EU FP6, and hosted in the cities of Amsterdam, Leipzig and Genoa, aimed to promote a user-friendly and user-relevant Information and Communication Technology (ICT) Platform that would offer integrated electronic services (e-services) and content for a plurality of users or actors (stakeholders) including tourists, residents, decision-makers, local managers, and civil servants. This article presents action carried out with Genovese students.

Research methodology

The research methodology of research chosen was the "Action-Research" approach (also known as Participatory Action Research, see Gilmore, 1986) with the addition of IT-based recursive cycles (Varisco, 2002). Recursive cycles allow a step-by-step monitoring of the project development while making revisions or adjustment during the implementation phase easier (Rocca, 2003).

Such an integrated approach, combining ICT tools with a participative methodology, allowed a geo-referenced Web system to be built as an instrument to facilitate sharing and communication among different actors (or stakeholders). The participatory processes were activated with groups of students by these interactions and are presented and analysed here in terms of procedural aspects and e-services decision-making.

Materials: The ISAAC tool

On the basis of the hypothesis that the planning and the structure of the methods are strategic to facilitate participation, a tool was created (http://www.feem-project.net/isaac/) reflecting the three directions of on-line participation mentioned earlier (OECD, 2003; Vedel, 2003; Macintosh, 2003).

The information section aimed to inform and sensitise people about local cultural heritage resources. Here, the user has a passive attitude as he just receives information and data without any interaction with other users or service providers. The information section was developed with the contribution of one school of the city of Genoa and the municipality of Genoa itself, as partners of the ISAAC project. It presented an overview of the main cultural sites and resources of the city, with all the relevant links to get more detailed historical and cultural data, as well as other information related to important events, news bulletin, facts, etc.

The communication section sought to create a virtual public space for exchanging general opinions and ideas about cultural heritage among citizens. Communication among students was established using on-line debates and surveys, with the possibility of uploading photos, images, videos and text documents. On-line debates gave the opportunity to discuss anonymously, to create a virtual dialogue with young people without having to meet them, to compare ideas and experiences, to experiment with new co-produced rules, based on mutual decisions and shared by the participants. In particular, in this section there were three links for communication: "your voice", "surveys" and "debates". "Your voice" was a repository collecting videos, photos, audio, and texts uploaded by the users, as a means of discovering and appreciating the city in an original way. Photos, stories, and videos therefore became a source of dialogue with other web citizens. Using the link "Your voice" it was possible to upload files that could be edited on the site. The link "Surveys" allows them to participate in on-line questionnaires[3]. Furthermore, participants had the possibility of inserting general comments, which could be read by all users. The link "Debates" allowed a dialogue among users to be established on proposed issues. A facilitator moderated debates, by asking the citizens specific questions on cultural heritage.[4] Students were asked to provide comments on the website, to identify which sections were the most appreciated, which parts should be improved and how.[5]

---

[3] The on-line surveys include specific questions related to cultural heritage in Genoa: *(i) Which are the objectives of a civic museum web portal in Genoa?*; *(ii) Which of the following web portals do you think offer more suitable on-line services?*; (*iii*) *Which of the following sites, monuments and areas do you think are the most relevant for the community of Genoa?* These questions have the general objective of encouraging people to provide suggestions about specific issues on cultural heritage. This section can be adapted in the future by including further surveys or different questions addressing issues of concern of the Municipality.

[4] Debates in this section have been moderated by a member of the research team of FEEM. Whenever this instrument is used in the future by the Municipality of Genoa, a specific person has to be instructed and charged to moderate discussion.

[5] Other questions have been inserted in the "debate" area with a general aim of encouraging a dialogue among citizens through the web. More specifically the following debates have been activated: (i) *Which parts of this web site do you think should be improved? How? What is still missing that you would like to find?*; *(ii) Do you think that citizens, tourists and service providers should be involved in realizing a website focusing on cultural tourism? Why?*; *(iii) What do you think are the most significant sites in the city of Genoa? Are these sites protected and enhanced enough?* Furthermore, specific debates have been activated on the arguments proposed in the on-line surveys.

In terms of participation, the tolls aimed at creating a more advanced level of interaction between students, tourists and service providers. It has been carried out using a geo-referenced tool based on e-blogs. This section was moderated by the research team with the main objective of identifying the e-services, which could be developed in order to improve access to cultural heritage and to enhance its resources. The discussion was geo-referenced, which means comments and suggestions inserted by people were immediately linked with a specific site in the city.

The tool was developed and tested according to Web 2.0 possibilities and following a participatory approach based on territorial mapping. In the Blended Focus Groups on-line activities, participants had been asked to indicate specifically which e-services they think should be developed for valorising cultural heritage in Genoa and for improving its accessibility. Responses were reported in blogs linked to the specific cultural sites identified by the participants as the most significant ones in the city. On the right hand side of the screen were the specific question about e-services identification and below the responses of the participants. A red marker linked to specific points in the map, allowed the participants' answers to be associated with a specific site, visualising it on the right hand side of the monitor. The process resulted in the construction of a cognitive thematic map showing the desired e-services to be developed for each identified site.

The participative section had thus been specifically designed in order to pursue this objective by involving all the relevant social actors. Nevertheless, it should be made clear that the participation section aimed at activating a dialogue with decision-makers (Public Administration), too: in this sense, it represents a useful tool for the Municipality of Genoa, which could be used to address other issues under concern after some suitable adaptation (for example cultural heritage access and management, social groups, etc.). The "participation" reserved area was structured using friendly and open source tools, and the dialogue activated using e-blogs associated to specific geographical sites in Genoa. The instrument allowed the user to contribute to the discussion by autonomously managing time and places for participation.

Blended Focus Group (BFG) through the GeoBlogs

The methodology employed in our case study required the establishment of a series of focus groups organised with a secondary students both face-to-face meetings and on-line discussions. This mixed or "blended" method (BFG: Blended Focus Groups) for debate and dialogue was particularly suitable for the complex and diversified situations such as those we live in today (Chiarullo and Rocca, 2003). It offers teachers the opportunity to manage times and places for students' autonomous participation, thus making it possible to adapt the approach to each person's needs. The proposed activities aim at giving a voice to the "broader" local community which extends to the network community using strategies to facilitate both face-to-face and on line participation (Chiarullo and Rocca, 2007).

The idea of BFGs took form from thoughts developed within the "Blended learning" view, a teaching method that integrates e-learning education with face-to-face meetings.[6] For our purposes, the intention was to offer the possibility to reflect about what emerged from the focus groups and to develop a personal contribution; at the same time, this possibility was also offered to people who could not intervene face-to-face (Rocca, 2005).

The face-to-face activities were meant to identify what the participants perceived as "territorial heritage" in the city. The on-line debate was organised in order to select the e-services that could be used in an integrated way to improve the accessibility to the previously identified city heritage. Participants were asked to enter the website (http://www.isaac-genovaculture.eu), which was then described to them by the facilitator by guided navigation. This session was organised in a room with PC facilities and a facilitator moderated the discussion. The students were asked to associate specific e-services (or integration of e-services) to each site that had been identified as most significant for the city in the face-to-face stage. The discussion was activated using the GeoBlog area, where participants had to report their answers. Before entering this reserved area, a Google Map frame was updated by identifying with a marker the sites, which had been declared as significant for Genoa in the previous face-to-face activities.

After the session, participants were asked to continue the discussions by accessing the on-line debate from home.

---

[6] This type of solution is easily adapted to the context of lifelong learning and to university teaching, which have high levels of complexity requiring flexible and open solutions, through the use of a broad spectrum of technological tools and teaching methods.

Face-to-face activities	1a) A *questionnaire* is administered to the participants who are asked to identify what they perceive as territorial heritage in Genoa and the added values of the territory .
	1b) A *working group discussion* follows, moderated by a facilitator: the students are asked to select, from the complete list of the sites they reported in the questionnaire (the most significant for the city of Genoa), and to identify their contribution to the community, the weaknesses and possible actions for improvement
	1c) The facilitator starts a working group discussion involving all the participants about some issues of interest.
on-line activities with GeoBlogs	2a) Participants are asked to enter the website, and the facilitator describes it with a guided navigation. Participants are then asked to start the on-line debate. This session is organised in a room with pc facilities and the discussion is moderated by a facilitator.
	2b) Participants are asked to continue the discussion started in the previous session by acceding to the on-line debate from their home; the discussion is moderated by a facilitator.

Figure 9.2. BFG activities.

Figure 9.2. summarises the BFG activities. It should be noticed that sessions 1a), 1b) and 2a) were organised using researcher facilities or other places where the participants were invited. Session 2b) was managed by the individual him/herself who could take part in the debate from his/her home. The on-line activities were not necessarily done after the face-to-face activities, but they could be activated during the first step.

Figure 9.2. shows the methodological framework, which was followed to structure the web tool and to identify the e-services for the territory: the process (creation of the website and participation tool), the method (blended focus groups and participation tool), and the products obtained (identification of sites of interest and e-services to be attached). A discussion among students was activated using the GeoBlog area of the website where each of the participants inserted their responses (in terms of e-services) which, in turn, were visible to the others.

The GeoBlog was structured to create a participation interface. The objective was to have a virtual space in which all the proposed ideas could be discussed and compared providing a new and wide source of "low cost knowledge" to the decision-makers (and the Public Administration).

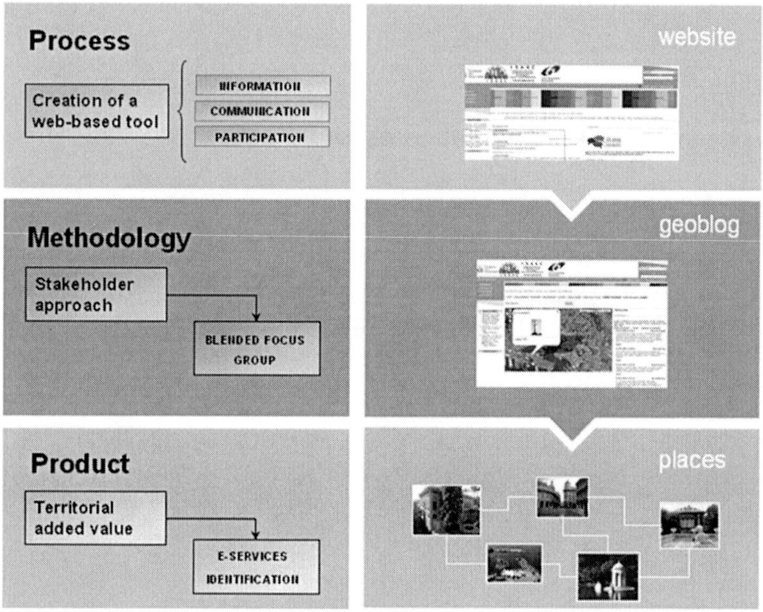

Figure 9.3. Process, methodology, product.

As a result of this approach, the participants were able to express their opinions and as a result they were able to add value to the resources of the region (Rocca, 2010; Rocca, 2007; Chiarullo, 2007). In consequence, the overall territorial value was increased. This approach aimed at capturing a wide range of opinions; therefore, three different groups of students were involved as they experienced the territorial heritage of the city in different ways. As was anticipated, the mixed or "blended" method (BFG) required the use of focus groups.

In the ISAAC project three different groups of social actors were involved: young residents, young tourists and service providers. In this case, the residents were concretely "living" the city heritage in their everyday life and were therefore considered important actors to suggest actions in order to improve access to these resources. Tourists were included for their potential role in cultural tourism development to improve the tourist offer. Finally, service providers were qualified experts on the service provision at the city level and were contacted to provide a specific contribution on how to enhance city heritage from the perspective of the supplier. Both local and external service providers were involved. The first were local agents, such as the municipality, local public transport, tourist agencies, cultural associations, museums, and so on. The second were external agents with a competence not limited to the specific territory of Genoa, which was judged useful in order to promote successful ideas that had already been put in place in other cities (e.g. associations for the conservation of historical and cultural heritage, and IT providers).

More specifically, in this project the first two focus groups met in April-May 2007. The first was addressed to young residents chosen in the secondary school (8 participants), while the second was to young tourists (8 participants). The two other focus groups met in October-November 2007, the first addressed to local service providers (11 participants) and the second to external service providers (10 participants). Thus these four focus groups involved 37 participants altogether, who had been recruited by FEEM Culture Factory premises located in Genoa. Participants had first been contacted by sending a letter of invitation, and then by phone; at the end of the focus group activities, participants received an oil coupon of 20 Euros. Furthermore, young residents and young tourists received free tickets for four civic museums in Genoa, offered by the Municipality.

Results of the participatory process

It is possible to outline two outcomes activated by the participation process. The first was a "product" outcome, referred to the list of the most significant places in Genoa and of the e-services young people had identified for each place; actors were in fact asked to identify the most important places in terms of territorial value and to spell out the most suitable technological solutions to be associated to the places in terms of e-services supply. The second is a "process" outcome, linked to the efficacy of the methodologies. The latter was monitored with the SERVQUAL methodology, which was used to test the website usability and functionality and to identify criteria and features judged as important by the users (for the results, see Rocca, et al., in press). On the other side, it allowed a constant monitoring of citizens' expectations and satisfaction with decision-makers and service providers. In this sense, the tool could be used as a satisfaction analysis assessment module. In a future development it could be used as a permanent instrument for monitoring people's expectations and satisfaction (Chiarullo and Rocca, 2007). The focus of the remainder of this contribution will be on the "product" outcome.

In terms of product, the project produced a series of maps highlighting on the one hand the places recognised as most valuable in terms of territorial importance and, on the other hand, for each of the mapped places, specific e-services identified by involved actors to increase accessibility and to improve visitors' experiences. The actors involved in the project believed that e-services would lead to better accessibility to local heritage and, as a consequence, a more enjoyable experience. As a first result of this process, a number of sites were identified in Genoa as part of the perceived heritage of the city, as reported in Table 2. They included cultural sites and cultural heritage, social sites and environmental sites (Chiabai et al, 2008; 2011).

Maps about territorial values underlined how places, which are generally not so popular among tourists, actually have a significant, special value for local actors. They are of special importance and a sense of originality and uniqueness is attached to these places. It is worth commenting on the number of sites and their variety, which was not limited to the traditional tourist itineraries.

Sites	**Examples**
Cultural sites and cultural heritage	Old Town, Garibaldi Street and Rolli Palaces, San Lorenzo Church, Ducale Palace, walls and fortresses
Social sites	Ancient port, Piazza delle Erbe, Aquarium
Environmental sites	Lanterna, Boccadasse and Nervi Gardens, walls and fortresses

Figure 9.4. City heritage perceived as the most significant in Genoa. (Adapted from Chiabai et al, 2011).

The results of this process concerned the e-services and how their proposed integration for each site would improve its accessibility. The results varied according to the specific site and heritage under analysis, and depended on the type of actor addressed. In general, "Information Services",[7] "Profiling", and "Interactive map with virtual tours" were considered as essential services and have been mentioned for all the sites. These services are often associated with specific profiling systems to avoid an overload of information. While a GeoBlog was considered a very

[7] Web pages including textual information and data (historical-cultural, tourist, etc) linked to a specific site of the city. They also include an event calendar and the possibility of thematic searching.

important support tool for providing the above-mentioned e-services, the most mentioned tool was the mobile phone. "E-Participation"[8] and "e-Governance"[9] services were more relevant to residents than tourists, but they were not frequently mentioned with the exception of the Old Town location.

The results obtained for one particular site will be discussed more specifically in order to illustrate the benefits of this approach and take into account the answers given by the different actors: the Old Town has been chosen, as it represents a key site for the city (cultural heritage enjoyment, tourism activities, transport, etc.) and because of the variety of e-services (and of their integration) that were proposed in this context. Furthermore, most of the participants focused on the Old Town, which confirmed the relevance of this site.

In the Old Town, young residents suggested a package of integrated e-services, which we have termed the "City Notice Board" (Figure 9.5.), structured on different themes using personal profiling and including information services, newsletter, event calendars, e-communities and e-governance services. These latter were considered key services for the Old Town as they provide the opportunity to reach a shared vision about the actions to be taken in this area of the city, characterised by a high number of problems and issues. Participants thought that e-governance tools should be used in order to propose ideas and suggestions about city sites management, to interact with other citizens about events, to monitor the necessary actions, to enhance the site and to improve decision-making. They appreciated these instruments as a tool for communicating their opinions and for interacting with the municipality. It was suggested to develop the "City Notice Board" within a specific website structured for this purpose.

---

[8] Using web forums, chat rooms, messaging programs and the like, it is possible to exchange information and compare ideas with other users, and to make suggestions about ways to contribute to the cultural life of the city

[9] A Web system that facilitates interaction among citizens and local authorities. It provides access to virtual city offices and databases. In some cases these systems make it possible to vote using the internet and to participate in on-line opinion polls.

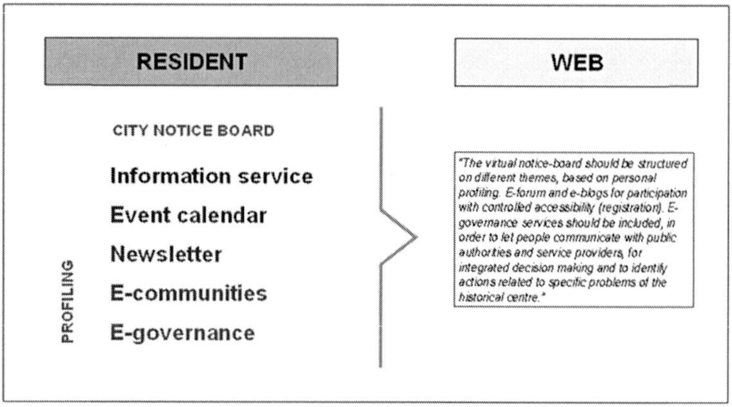

Figure 9.5. e-services for young Residents.

As far as young tourists are concerned, a "City Tour" package of integrated e-services was proposed (Figure 9.6.), which includes information services, interactive maps, journey planners[10] and virtual tours, all based on personal profiling.

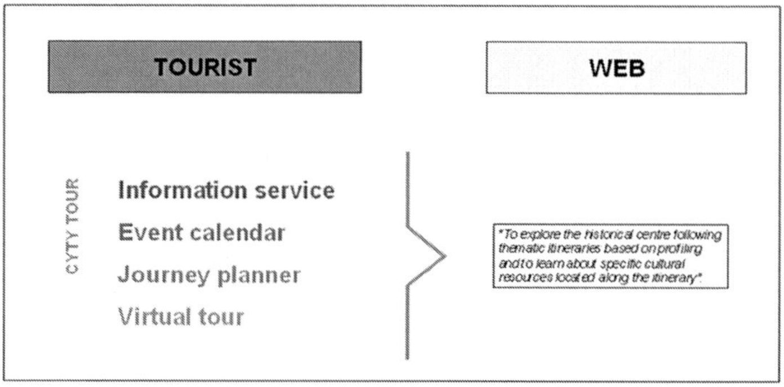

Figure 9.6. e-services for young Tourists.

The purpose of e-services for young tourists was exclusively orientated towards visiting and enjoying the city from a cultural, historical and tourist

[10] A set of services that makes it possible to plan a trip to any tourist destination. It is also possible to download maps, guided tours, information, and advice about the length of stay and means of transport onto one's own machine (pc, mobile phone, palm).

point of view. The proposed "City Tour" should be developed within a specific website. In addition, kiosks (or e-centres) were suggested as important support for the e-services, allowing access to interactive maps and virtual tours, and the possibility to customise visits in the Old Town. Finally, tourists also proposed the use of the mobile as a means of providing information services and interactive maps based on personal profiling (Figure 9.7.).

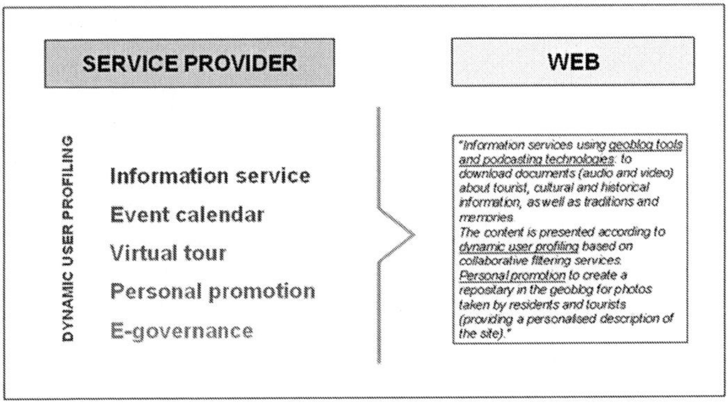

Figure 9.7. e-services for Service Provider.

Informative e-services were considered essential to better appreciate the site and to find useful news about the young tourist sites and heritage existing in the area of interest; to receive updating about programmed events (newsletter and promotional messages); and to receive suggestions about tours and itineraries (kiosks and virtual tours).

Finally, service providers suggested a more sophisticated package of integrated e-services (Figure 9.8.), for "Dynamic City Use", including information services, an event calendar, interactive maps, virtual tour, personal promotion and e-governance services. The first four services should be integrated within a dynamic user profiling system. Dynamic profiling allows the user's individual demands to be met by collecting information about many other users' experiences and preferences.

Another set of considerations was related to the link between e-services and a single place, as specified by the BFG participants. In particular, Figure 9.8. presents the specific e-services suggested by each social actor (residents, tourist and service providers) for each of the sites identified as significant for Genoa. The results varied according to the specific cultural site under analysis. In general, for all the sites, the e-services with the highest priorities tended to be "web information" and "interactive maps with virtual tours" (resulting as the first or the second priority depending on the site concerned). E-governance services were mentioned only for the “Old Town”, due to its territorial specificities characterised by many activities and functions.

e-services suggested for the Old Town

	Young Resident (web site)	Young Tourist (web site and mobile)	Service providers (web site and mobile)
e-services	Web Information	Web Information	Web Information
	Profiling	Profiling	Profiling
	Newsletter	Video, photo and audio downloading	Video, photo and audio downloading
	e-communities	Journey planner	***Interactive map with virtual tours***
	e-participation	***Interactive map with virtual tours***	e-governance
	e-governance	Blogs and personal promotion	---

e-services suggested for the Ancient Port

	Young Resident	Young Tourist (web site and mobile)	Service providers (web site)
e-services	*No suggestions*[xi]	Web Information	Web Information
		Newsletter	Profiling
		Video, photo and audio downloading	Video, photo and audio downloading
		Virtual games	***Interactive map with virtual tours***
		Promotional spot	e-participation
		Interactive map with virtual tours	e-governance
		e-communities	---

e-services suggested for Via Garibaldi

	Young Resident (web site and mobile)	Young Tourist (web site and mobile)	Service providers (web site and mobile)
e-services	Web Information	Web Information	Web Information
	Virtual games	Newsletter	Profiling
	One stop shop	Video, photo and audio downloading	Video, photo and audio downloading
	Promotional spot	One stop shop	Promotional spot
	e-communities	Blogs and personal promotion	***Interactive map with virtual tours***

e-services suggested for San Lorenzo Church

	Young Resident (web site)	Young Tourist (web site)	Service providers (web site)
e-services	Web Information	Web Information	Web Information
	Profiling	Virtual games	Video, photo and audio downloading
	Newsletter	---	Journey planner
	Journey planner	---	***Interactive map with virtual tours***
	Interactive map with virtual tours	---	e-participation

e-services suggested for Lanterna

	Young Resident (web site)	Young Tourist	Service providers (web site and mobile)
e-services	Web Information	*No suggestions*[xii]	Web Information
	Profiling		Newsletter
	Newsletter		Video, photo and audio downloading
	Video, photo and audio downloading		***Interactive map with virtual tours***
	One stop shop		e-governance
	Interactive map with virtual tours		---
	e-participation		---

e-services suggested for Boccadasse

	Young Resident	Young Tourist (web site)	Service providers (web site)
e-services	*No suggestions*[xiii]	Web Information	Web Information
		Newsletter	Video, photo and audio downloading
		Video, photo and audio downloading	***Interactive map with virtual tours***
		Interactive map with virtual tours	---

e-services suggested for Nervi Gardens

	Young Resident (web site and mobile)	Young Tourist (web site)	Service providers (web site and mobile)
e-services	Web Information	Web Information	Web Information
	Promotional spot	Newsletter	Profiling
	One stop shop	Video, photo and audio downloading	Video, photo and audio downloading
	Interactive map with virtual tours	***Interactive map with virtual tours***	Promotional spot
	e-communities	Blogs and personal promotion	***Interactive map with virtual tours***
	e-governance	---	---

Figure 9.8. e-services suggested to improve access to cultural heritage.

Discussion

Even if a web-based device was created, paper-based approaches were employed in our case study, as an irreplaceable instrument of mediation. In the geo-referenced activities, the use of paper allowed participants to locate their thoughts in a spatial order and in a stable way. The signs that the users decided to draw on the maps helped creativity and suggested new ideas and solutions. New itineraries, and the discovery of different views (linked to personal and territorial experiences) were possible (Amoretti, 2007). This is coherent with the approach of Participatory GIS (Brown, 2012; Dragićević and Balram, 2004) but employs low cost tools, which do not require specific knowledge. The use of cartographic representation activates a process that, from the codified image of the world (map), can generate images and new visions, which are in their turn represented on the map, in a sort of circular movement made up of feedbacks and steps forwards. Once again, virtual reality and the real world can be closely linked (Farinelli, 2008; Tang and Waters, 2005). Collaboration is therefore facilitated, as different young people are able to talk, confront, and exchange ideas. The diversity of contexts and way of thinking favours a real change in both individual and collective behaviours.

It has to be mentioned that according to González, Gilmer, Foley, Sweeney and Fry (2008):

- nobody can make changes alone: if a change was made by someone, the entire network of their relationships would be affected by it;
- the network of relationships created by individuals is a collective actor characterised by common values, established objectives, convergent interests, integrated behaviours, participated sentiments, operative practices, distributed responsibilities;
- the network is therefore a communication network: actors are connected by a flow of information, so that the system can act as an identified and singular subject; and
- the network uses a communication system that broadcasts information, makes messages more effective, and reaches intelligent terminals (the people forming the network).

Consequently, participation, communication, and resource enhancement are the conditions that allow accountability in a project as the one we have described.

The geo-referenced website was constructed to create a participation interface (Robinson, Eslambolchilar and Jones, 2008). The objective was to create a virtual space in which the ideas are discussed and compared providing a new and wide source of low cost "knowledge" to the decision-makers (and to the Public Administration). This instrument was a clear and transparent tool for residents and tourists to make suggestions and send their opinions and requirements. The idea underlying the construction of the tool was that people living and experiencing the area are those who can really identify sites of interest in the city and the e-services, which can help to improve their experience.

The map of the Genovese region was constructed in a cooperative/collaborative way to show the most relevant resources there and associated e-services to improve accessibility according to the viewpoint of local young actors, decision makers and young tourists who become protagonists in the definition of the visibility of the resources. The e-services which were mentioned may appear trivial; actually, the choices which were made demonstrated little knowledge of technologies, but it was clear that people have chosen what they think is closer to them.

The technique includes the use of cartographic representations/constructions that made it possible to locate things, movements and also thoughts – in a spatial and stable order. However, the marks on the paper that give certainty to the elements they represent, suggest at the same time other things leading to the discovery of a new order and new significance of things, while revealing unknown and exciting itineraries. The same "territorial objects", once they have been mapped, leave a margin for interpretation and can be configured in different ways according to different points of view. As Dematteis (2008) observes, if on the one hand cartography defines "where", based on what is already known, on the other hand it makes people imagine new forms and interpretations of the world that surrounds them. The process used to create the cartographic representation is thus a continuous circular process: the codified image of the real world generates new visions and representations, which can afterwards emerge and crystallise. The creative activity that underlies this representation is unquestionable. This grassroots process has several advantages over constructing a territorial identity since mapping the area with the citizens will be the product of a participatory process, able to catalyze resources and energies among the people involved. The reality of "public property" like a city seems to derive not from marketing actions, but rather from the value assigned by citizens' affection and empathy: only undergoing this stage it can become a heritage.

The methodology used in this research could be replicated in other cities and contexts, but it would require the development of a new website and the creation of new focus groups, following the steps discussed in the previous sections. For the city of Genoa, however, the website is a tool which can be applied in the context of city heritage management for discussing issues related to territorial sites conservation.

The case study of Genoa, as discussed in this chapter, presents a number of caveats. Each group of young participants involved in the participatory process consisted of a limited number of actors (8 to 10). It must be noted however that the exercise is not meant to be a quantitative survey representative of all the young population of young tourists, young residents and service providers in Genoa. The methodology used, the blended focus groups, is a qualitative survey methodology which usually involves a limited number of people who are asked to discuss some themes in detail during one or two hours. The objective was not to derive a representative sample and provide figures that are representative from a statistical point of view. The qualitative discussion between participants was the purpose of the method, as the themes and questions presented are complex and multifaceted, and therefore quantitative surveys are not, from our point of view, the appropriate methodology to use in this context. From this perspective, we cannot argue that the results are representative of all the population in the city of Genoa, but they do provide some qualitative inputs for the construction of integrated packages of e-services meant to improve accessibility to city heritage. Once the main features of e-services packages are decided, a further step could be the administration of a larger representative survey to define sub-attributes and minor characteristics of the services.

A second remark that should be made is that the young participants invited to the focus groups had a minimum knowledge of computer and ICT tools. This was in fact a pre-requisite for participation, otherwise a learning process would have been required as a first step and this was outside the scope and length of the project. However, the issue is relevant as it denotes the problem of the "digital divide" which refers to existing disparities between population groups having and not having access to, and knowledge of, information technology. Future developments of the method should take into account this issue by adapting the participatory process for people not using ICT.

Acknowledgments

The study was part of the European 'Framework Sixth' Project ISAAC: "Integrated e-Services for Advanced Access to Heritage in Cultural Tourist Destinations", a multi-disciplinary research initiative aiming to enhance cultural assets as tourism resources through user-friendly and stakeholder-relevant integrated e-services in urban tourist destinations. Our thanks go to Aline Chiabai for sharing the project; and to Cristina Minelle for the translation.

References

Alford, P.; Open Space – a collaborative process for facilitating tourism IT partnerships; in O'Connor, P., Höpken, W. and Gretzel, U. (Eds.): *Information and Communication Technologies in Tourism*, Springer-Verlag, Wien, 2008; pp.430-440.

Amoretti, F.; International organizations ICTs policies: e-democracy and e-government for political development, Review of Policy Research, Vol. 24, No. 4, 2007; pp.331-344.

Anthony A, Atkinson, J., Waterhouse H., Wells R. B.; A Stakeholder Approach to Strategic Performance Measurement; Sloan Management Review, Vol. 38, Issue 3, 1997; pp. 25-37

Armstrong, C.; Emergent democracy. In *Open Government* Lathrop D., Ruma L., Eds.; O'Reilly, Sebastopol, CA USA, 2010.

Asgarkhani, M.; The effectiveness of e-service in local government: a case study; *The Electronic Journal of e-Government*, Vol. 3, No. 4, 2005; pp.157-166.

Bramwell, B. and Lane, B.; Collaboration and partnership in tourism planning; in Bramwell, B. and Lane, B. (Eds.); *Tourism Collaboration and Partnerships: Politics, Practice and Sustainability*, Channel View Publications, Clevedon, UK, 2000; pp.1-19.

Brown, G, A; Empirical evaluation of the spatial accuracy of public participation GIS (PPGIS) data, *Applied Geography*, Vol. 34, May 2012, pp.289-294.

Buhalis, D. and Pistidda, L.; The impact of WiMAX on Tourist Destinations; in O'Connor, P., Höpken, W. and Gretzel, U. (Eds.); *Information and Communication Technologies in Tourism*; Springer-Verlag, Wien, 2008; pp.383-394.

Buhalis, D.; *eTourism: Information Technology for Strategic Tourism Management*; Pearson Education Limited, Harlow, 2003.

Carter, L. and Belanger, F.; *The utilization of e-government services: citizen trust, innovation and acceptance factors*; Information Systems Journal, Vol. 15, 2005; pp.5-25.

Carugati A., Hadzilias E. and Demoulin, N.; *Setting the Framework for Developing eGovernment Services on Cultural Heritage*; ECIS 2005 Proceedings, 2005; Paper 132.

Chiabai, A., Lombardi, P., Chiarullo, L., Rocca, L., Paskaleva-Shapira, K., Brancia, A.; *An e-governance system for managing cultural heritage in urban tourist destinations: The case of Genoa.* In Collaboration and the Knowledge Economy: Issues, Applications, Case studies, Cunningham, P., Cunningham M. (Eds). IOS Press: Amsterdam, 2008.

Chiabai, A, Paskaleva-Shapira, K., Lombardi, P.; E-participation model of sustainable cultural tourism management: a bottom-up approach. *International Journal of Tourism Research*, published online in Wiley Online Library, DOI: 10.1002/jtr.871, 2011.

Chiarullo, L.; *E-Governance e E-Participation: dalla teoria alle pratiche*; Quaderno sullo sviluppo sostenibile, 2, FEEM, Milano (2007).

Chiarullo L., Rocca, L.; *E-Governance e Autososenibilità Locale. Rapporto per lo Sviluppo Sostenibile*, Fondazione Eni Enrico Mattei, Milano, 2007.

Chiarullo L., Rocca, L.; *Sistema PANDORA. Progetto per Venezia.* Equilibri, 3, Il Mulino, Bologna, pp. 273-290, 2003.

Ciborra C.; *The Labyrinths of Information: Challenging the Wisdom of Systems*; Oxford University Press, Oxford, UK, 2002.

Costa N.; *I professionisti dello sviluppo turistico locale*; Hoepli, Milano, 2005.

—. *La città ospitale. Come avviare un Sistema turistico locale di successo*, Mondatori, Milano, 2008.

Dematteis G.; Luoghi vissuti, luoghi inventati: la diversità geografico-culturale, in M. Bertoncin e A. Pase (a cura di*), Pre-visioni di territori. Rappresentazioni di scenari territoriali*, Franco Angeli, Milano, pp. 54-70, 2008.

Dragićević, S, Balram, S.; A Web GIS collaborative framework to structure and manage distributed planning processes. In: *Journal of Geographical Systems*, Vol. 6, Issue 2, pp. 133-153. Berlin, 2004.

Eaves, D.; After the collapse: open government and the future of civil service; In *Open Government*, Lathrop D., Ruma L., Eds.; O'Reilly, Sebastopol, CA USA, 2010.

European Commission; *Public Participation in relation to the Water Framework Directive*, European Communities, 2003.

Farinelli, F.; *Che cos'è il territorio (e perché crediamo alle mappe). Previsioni di territori. Rappresentazioni di scenari territoriali*, in M. Bertoncin e A. Pase (a cura di), Pre-visioni di territorio, Milano, Franco Angeli, pp. 21-40, 2008.

Freeman, R. E. and McVea, J.; *A Stakeholder Approach to Strategic Management*. Darden Business School Working Paper No. 01-02. Available at SSRN: http://ssrn.com/abstract=263511, 2001.

Freeman, R.E.; A Stakeholder Approach to Strategic Management, eds. M. A. Hitt, R. E. Freeman, & J. S. Harrison, *Analysis*, 1(01), 1984, p.276. Available at: http://books.google.com/books?hl=en&lr=&id=zD1CZUWE6zQC&oi=fnd&pg=PA189&dq=A+Stakeholder+Approach. (accessed on 21/02/2013)

Gilmore, T., Krantz J. & Ramirez R.; *Action Based Modes of Inquiry and the Host-Researcher Relationship*. Consultation, 5 (3), 1986, pp. 43-56.

González, A, Gilmer, A., Foley, R, Sweeney, J, Fry, J.; Technology-aided participative methods in environmental assessment: An international perspective, *Computers, Environment and Urban Systems*, Volume 32, Issue 4, July 2008, pp. 303-316.

Hanzl, M.; Information technology as a tool for public participation in urban planning: a review of experiments and potentials, *Design Studies*, Volume 28, Issue 3, May 2007, pp. 289-307.

Kalay Y., Kvan T., Affleck J.; *New Heritage New Media and Cultural Heritage*, Routledge, 2007.

Lacoste, Y.; *La géographie, ça sert, d'abord, à faire la guerre*, Paris, Maspero, 1976.

Mitsche, N., Reino, S., Knox, D. and Bauernfeind, U.; Enhancing cultural tourism e-services through heritage interpretation; in O'Connor, P., Höpken, W. and Gretzel, U. (Eds.): *Information and Communication Technologies in Tourism*, Springer-Verlag, Wien, 2008; pp.418-429.

Macintosh, A.; Using information and communication technologies to enhance citizen engagement in the policy process, in OECD, *Promises and Problems of E-Democracy: Challenges of Online Citizen Engagement*, Paris. 2003.

OECD; *Engaging Citizens Online for Better Policy-making*, Paris, 2003.

Robinson, S., Eslambolchilar, P., Jones, M.; *Point-to-GeoBlog: gestures and sensors to support user generated content creation*, in Proceeding MobileHCI '08 Proceedings of the 10th international conference on Human computer interaction with mobile devices and services, New York, NY, USA 2008, pp. 197-206.

Rocca, L.; *Il territorio della rete*, Lecce, Pensa Multimedia, 2003.

—. *Opening PANDORA's Box: Participatory Network for Venice's Sustainability*, in FEEM Newsletter, 1, FEEM, Milano, 2005; pp. 24-30.

—. *Partecipazione come pratica territoriale*, Equilibri, Milano, 1, 2007; pp. 117-124.

—. *Partecipare in Rete, Nuove Pratiche per lo Sviluppo Locale e la Gestione del Territorio*; Il Mulino, Bologna, Italy, 2010.

Schegg, R., Liebrich, A., Scaglione, M. and Syed-Ahmad, S.F.; An exploratory field study of Web 2.0 in tourism, in O'Connor, P., Höpken, W. and Gretzel, U. (Eds.): *Information and Communication Technologies in Tourism*, Springer-Verlag, Wien, 2008; pp.152-163.

Tang, K. X., Waters N, M.; The internet, GIS and public participation in transportation planning, *Progress in Planning*, Volume 64, Issue 1, July 2005, pp . 7-62.

Tuan, Yi-Fu; *Human Goodness*, University of Wisconsin Press, Madison, 2008.

Varisco, B. M.; *Costruttivismo socio-culturale. Genesi filosofiche, sviluppi psico-pedagogici, applicazioni didattiche*, Carocci, Roma, 2002.

Vedel, T.; L'idée de démocratie électronique: origines, visions, questions, in P. Perrineau (eds.), *Le désenchantement démocratique*, La Tour d'Aigues, Editions de l'Aube, 2000.

CHAPTER TEN

INTRODUCING GIS IN GREEK COMPULSORY SCHOOLS: VISION OR REALITY?

AIKATERINI KLONARI

Introduction

Geographic Information Systems (GIS) is a set of integrated software programmes designed to store, retrieve, manipulate, analyse and display geographical data-information concerning people, places and the environment. GIS has emerged in the last decade as an essential tool that plays a key role in human activities in everyday life. It is one of the fastest growing uses of computer technologies and is a fundamental part of modern geography (Koutsopoulos 2005). GIS is a tool that is being used extensively by researchers, scientists and administrators to inform decision-making about real issues (Hwang 2006). Additionally, GIS has been utilised in different disciplines, especially in geography at university level, and recently in teaching and learning different subjects in schools (Bednarz 2004).

As mentioned, the use of GIS started at an academic level (in American and Canadian universities) in the early 1980s and today constitutes an integral part of the curricula of all higher education geography departments worldwide. This fact somehow influenced the reforms of curricula in secondary education, making apparent the necessity of introduction GIS to schools, initially in a few and later on in many countries, in order to motivate students to use new technologies not only in the school environment but also in their everyday life (West 2003; Patterson et al. 2003).

Since the 1990s, interest in that field has increased, and many studies about the application and effectiveness of teaching with GIS in secondary schools have taken place. Research revealed that GIS is in fact a very important educational tool, which helps inquiry learning, learning with problem solving, and which can be combined with many innovative practices, such as field work and project methodology. Due to these advantages, GIS is proposed to be included in the curricula of several subjects in secondary education, apart from geography, including such disciplines as biology, mathematics, economics, social and environmental studies, etc.

GIS has a number of advantages, particularly for geographic education, because it greatly contributes to the development of students' spatial thinking skills (Bednarz 2004; Houtsonen 2006; Johansson 2006; Kerski 2009). Additionally, GIS has simplified the processes of analysis and presentation of geographic information and it has increased the use geographic inquiry in the classroom. Many studies reveal that GIS is an effective tool in promoting students' geographic skills, including helping them to think spatially, analyse and make inferences about spatial data, as well as provide student and teacher motivation, and adopting project-based teaching and learning (Shin 2006).

According to Bednarz and Van der Schee (2006), geography teachers use GIS for three main reasons: 1) GIS supports teaching and learning in geography; 2) GIS is a tool for investigating geographical problems of various scales; and 3) GIS constitutes an essential tool for the labour market and employability in the 21st century. Despite the fact that researchers argued that the use of GIS for both students and teachers has a lot of benefits, their integration in secondary education still remains a challenge.

On the other hand, research also indicates that the improper use of GIS in schools could have negative aspects, for instance if GIS were considered by students to be a "black box" that contributes nothing to the development of spatial thinking, geographic skills, understanding and problem solving in daily life. Therefore, in order to develop the proper use of GIS in class, it is paramount to maintain the relationship and boundaries between GIS and geographic education. Sui (1995) and Koutsopoulos (2010) propose two different aspects of integration of GIS in education: "Teaching with GIS" and "Teaching about GIS". The goal of "Teaching with GIS" is to allow students to study geography and develop geographic skills with the aid of GIS as a tool. The goal of "Teaching about GIS" is mainly to teach GIS technologies. According to them, the opinion that finally prevails with teachers with regards to the type of use of GIS in geographic education focuses on "Teaching with GIS". However, Johansson (2006) mentions that it is important to re-establish "Teaching about GIS" to some extent, before the use of GIS in class.

Applications of GIS in secondary education

Nowadays, despite the multitude of applications of GIS in secondary education, its use is not widely spread throughout the world (Kerski 2009). Even though some teachers consider GIS a potential means of reform, very few have, in fact, adopted it. Nevertheless, even today the reasons for the slow implementation of GIS in schools is related to the low levels of interest by education stakeholders and also that the level of its effectiveness in teaching and learning remains ambiguous. According to research, however (Olsen, 2002; Bednarz & Van der Schee, 2006; Kerski 2009), the key obstacles for the use of GIS in secondary education have been linked mainly to:

1) Technical aspects, such as the availability of hardware, software and data;
2) The lack of training of teachers and suitable instructive material;
3) Other issues that encourage or discourage innovation in education;

Certain additional obstacles reported in other studies include the lack of time for teachers to learn about GIS and to use it in activities in their classes, the unwillingness of teachers to learn the use of new technologies, and the insufficient mention of GIS in the curricula. The difficulty of using GIS software has also been reported as an obstacle. According to Bednarz and Van der Schee (2006), GIS software implies high technical requirements, and is a challenge to utilise since it was not originally designed for teaching/learning, and a lot of teachers do not know how or when to use it in class. Despite all these possible barriers, there has lately been increased interest in both the international community (Yap et al. 2008) and even the European Union for the integration of GIS in the educational process. This is apparent both from the attempts to simplify the use of GIS and from the creation of educational material and data for applications in schools, in the USA and Europe, from funded programmes. Moreover, certain countries that had already integrated GIS (United Kingdom for over a decade, Austria, Netherlands and Belgium) or recently included them in their curricula, such as France, Sweden, Finland and Turkey, have faced difficulties in the incorporation of GIS in the teaching of geography (Demirci, 2008). However, there are still not enough empirical studies related to the obstacles and barriers to the use of GIS in the secondary education in other European countries and, of course, in Greece.

GIS in Greek education

In Greece, GIS was introduced in universities during the 1990s predominantly at a minimal level. Nowadays, the number of GIS programmes available in tertiary education (universities and polytechnic schools) is satisfactory and constitutes an integral part of many departments' curricula. Moreover, in some of these departments, special courses of geo-informatics are provided at graduate, post-graduate and doctoral level.

On the contrary, until today, GIS did not appear (i.e. there is no direct reference) as part of the formal curricula in secondary education (either general or technical). It could be assumed that teachers had the opportunity to refer to geo-informatics through curricula available at school, if they were familiar with it. There is only a single reference to GIS in the book of Geology-Geography of 1st grade of Junior High school (Pavlopoulos & Galani 2009). If we take into consideration the particular importance that has been given in the past few years in Information and Communications Technologies (ICT) in secondary education and the training of teachers, we would expect to find some reference or application of GIS there, even at a basic level, in some of these courses. Furthermore, the few research studies carried out in Greece with regards to the use of GIS in secondary education (Kimionis 1995; Klonari 2009; Kontosi 2007; Klonari & Laina 2010) show that not only does GIS not exist in schools and that, naturally, it is not utilised in teaching and learning, but it still constitutes a "black box" for the overwhelming majority of teachers, even if there have been a significant percentage of science teachers that positively expressed their opinions on the prospect of using GIS in teaching geography courses.

Nevertheless, in Greece in the current decade, two European projects that related to the use of Geographic Information Systems in education were successfully completed. These were the three-year project GISAS (GIS Applications in Geography in Schools) that was completed in 2006. It was an educational and research programmes that was funded by the MINERVA action of the European Commission, with 8 European schools participating, among which one from Greece (2nd Senior High school of Larisa). Furthermore, the two-year project (2008–2010) IGUESS (Introducing GIS Use in Education in Several Subjects) was undertaken. This programme was a Comenius Project also funded by the EU, with the objective of training teachers and creating educational material which will facilitate and support the integration of GIS in education in several subjects (in addition to geography). In this project eight European countries participated, Greece among them (University of the Æegean). With the above-mentioned as an opportunity, and through this participation in the research programme IGUESS, two central questions arose: a) how and to what extent could GIS be incorporated in teaching and learning in secondary education in Greece, knowing that there is a deficit in relevant experience and taking into consideration the difficulties still faced by other, more experienced countries (United Kingdom, USA, etc.) towards integrating it in class; and b) how could GIS be assimilated and utilised by teachers in geography—let alone other courses—when they possessed limited knowledge of information technologies and GIS, but also of using open-source software and databases.

The iGuess Programme: a pilot course on Lesvos Island

In the past few years there has been a significant effort in Greece to improve the quality of education, through the new national curricula in all courses, by the creation of modern educational materials and various teacher training programmes, with the great majority of them using ICT. Despite all these efforts, schools today still face difficulties in the application of innovative programmes, due to insufficient infrastructure and funds, as well as the ability of teachers to use these new technologies in their courses. These are only some of the obstacles that restrict teaching in school from being student-orientated, focused towards methodologies of research and problem solving, as mentioned in the recent curricula (2003), and as Rocard et al. (2007) suggest to the European Commission.

Given this framework, it was not straightforward to utilise a new technological system such as GIS and to test its effectiveness in students' learning in several subjects or even in geography, since even today it is well known that teaching is predominantly teacher-orientated and the text book is the only resource used by students and teachers alike (Klonari & Koutsopoulos 2005; Germanos 2005; Rellou & Lambrinos 2008).

For the realisation of this research we faced the following initial problems:

1. Teachers were unfamiliar with the use of GIS;
2. They were unsure as to how to integrate them in their courses;
3. They did not have the necessary infrastructure in their school, such as the appropriate hardware and software.

These problems were resolved as follows: i) initially, teachers independent of specialism were asked to participate voluntarily in a training course for the use of GIS in teaching and learning, under the European project iGuess; ii) by issuing free whole school licenses for

ArcGIS 9.3 software for one year (and in subsequent years, if they chose to use it) in their school; and iii) development of educational supporting materials and accompanying digital data on activities that they could use for their courses (not only in geography).

Process of research

The research was completed in three phases: During the initial phase (3 days, 9 a.m.—6 p.m.) in October 2011 the teachers were trained at the GeoInformatics Laboratory in Geography Department of the University of the Ægean, and an initial evaluation of the activities created by the members of the project took place (via a questionnaire). During the second phase, the success of the application of the programme was monitored in the schools by the volunteers. Finally, in the third phase, the results of the application to the students were evaluated (worksheets and evaluation sheets).

The teachers that participated in the initial phase of the research for the programme were 10 in total, of various specialties (three teachers of informatics and computer science, three science teachers, one of Greek Literature and one English Language teacher, one elementary school teacher and one kindergarten teacher, working at the Environmental Education Centre), from five schools from Lesvos (two Junior and three Senior High schools) and from the Environmental Education Centre in Asomatos of Lesvos, with over 10 years' experience.

All teachers were familiar with the use of computers and spoke English fluently. However, only two said that they were familiar with GIS, but although they had never used it before for teaching purposes. In the third phase, 45 2nd grade students at the Experimental Junior High School of the University of the Ægean participated.

Six lesson plans were used in the teachers' training pilot course, there were online folders with digital data, worksheets, info sheets for each lesson and a list with i-Notes, a handy how-to-do guide developed for ArcGIS version 9.3 software uses. In addition questionnaires (for teachers and students) with closed and open questions were used for the evaluation.

Aim of the research

The main objectives of this research were to:

- familiarise teachers with teaching and learning methods using geoinformatics;
- develop ICT skills using specific examples in several subjects;

- use innovative teaching methods, student-oriented, such as inquiry and problem-solving, for effective learning; and
- evaluate learning outcomes of GIS application in schools.

Research results

In the first phase of this research, which was based on the pilot teacher training course, a selection of exercises were used, six in total, of varying degree of difficulty relating to the use of GIS in several subjects. The course consisted of three parts: in the first part the teachers were given basic information about GIS and its use in everyday life. Moreover, they were presented with open source software and web sites where they could find data. Consequently in the second part, the teachers conducted activities with the use of ArcGIS 9.3 software and the instructions in the lessons' worksheets developed by the iGuess project participants. Finally in the third part the teachers presented the finalised exercises, discussed and evaluated both of the try-out courses and of the activities created by the participants.

The following can be concluded from the evaluation:

a) Everyone (10/10) stated that through this training course they developed some new ideas as to how to use and implement GIS in their courses. In fact, six out of ten declared that they could and indeed wanted to apply what they learned in their school;
b) The answers to the question as to how much this training satisfied their expectations were: More than expected! (5), Exactly what I expected (4), and I would like to learn something more than what I learned (1);
c) Concerning the duration of the course, half of the teachers considered it very short; while the other half found that the duration was precisely what was needed;
d) Two of the teachers considered the "exercises" quite difficult, while the rest considered them within their capabilities. Eight teachers found the material that was given to them explicit and well structured and that they could use it in their own classes and courses, while two observed that they would prefer some additional and more explicit instructions and information on subjects that were not covered by given exercises;
e) Nine out of ten teachers stated that they would try and create their own course material with the use of GIS, which they would also share with other teachers. Only one teacher declared that she could not, because she did not feel very certain about it.

The second phase of the research was to monitor the teachers in their schools and establish whether they tried to involve, introduce or integrate GIS in their courses and teaching. While the results from the evaluation of the seminar were very promising, there existed, in fact, many problems throughout the phase of application to the schools. Initially teachers serving in the Senior High schools did not use ArcGIS due to lack of time and relevance of the existing exercises with their teaching subjects in school. Only one informatics teacher used some of the exercises and presentations as an example of the application of ICT in real world situations to his students. The teachers serving at the Environmental Education Center did not apply the exercises directly to their students,; although throughout the year they created their own exercises in environmental education. Additionally, in 2012, one of these teachers, who had already set up her own data (using GPS) in the field, created a scenario with role-playing games and worksheets on an environmental project and implemented these activities using GIS (map creation) in elementary school pupils of the 5th and 6th grades. Finally, the teachers at the two Junior High schools attempted to use the two sample lessons about "Earthquakes in Greece" and "Volcanoes in Greece" that they had completed during the pilot course. They were able to use both the data and educational materials (info sheets, worksheets and evaluation sheets) in geography. Nevertheless, neither of them managed to realise the particular exercises at school, because the PCs available to the students did not meet the minimum requirements for running the ArcGIS 9.3 software. Thus it was proposed to carry out the particular lessons in the laboratory of geo-informatics in the geography department. However this scenario was not feasible for one of the schools because it was too far from the University laboratory, and the teacher could not allocate an entire day for this activity (they were not given authorisation). Consequently, only the teachers from the Experimental Junior High School of the University of the Ægean were able to apply the GIS instruction in their geography classes.

The third phase of the research was the follow-up application of GIS in geography lessons in the university laboratory, with the use of educational materials from the teachers and their students, and finally the evaluation of the lessons. Through the exercises of the students, the achievement of the lesson activities' aims, the necessary time required for completion of the activities, and the degree of students' satisfaction from this particular lesson was evaluated. The exercise completed with the use of GIS was "Earthquakes in Greece" and the students were handed printed worksheets for their activities. The main objectives were for the students were: i) to visualise the spatial distribution of earthquakes that occurred in Greece over a certain period of time; ii) to locate the formation of seismic zones according to the distribution of epicentres of earthquakes and to correlate the seismic activity patterns to tectonic plates boundaries and active faults; and finally, iii) that they should then be able to identify cities at risk from seismic activity.

The methodology followed was guided inquiry (step-by-step) and the class duration was 2 hours. A total of 45 students (22 females and 23 males) in total participated. Each student worked individually on a PC with ArcGIS 9.3 and a folder containing all relevant data and information for the completion of the activities. The teachers (as facilitators) would help the students if asked for. By the end of this lesson, all students had created their own maps, which they printed and kept. It should be mentioned that before the start of the course the students were handed a questionnaire that, apart from personal data, contained questions concerning their familiarity with the use of PCs and GIS, and also questions for evaluating their knowledge relevant to earthquakes. It was concluded from the answers to the survey that all students were familiar with the use of PCs, which they used on a day-to-day basis, but none of them knew anything about GIS nor had any of them used it before. As for the responses with regard to what they expected from the course, the majority of the students answered that they expected it to be different and more interesting.

Some of the questions that concerned knowledge relating to earthquakes were:

- *a) Where in Greece do most earthquakes typically occur and why?*
- *b) Where (spatial distribution) in Greece, would you suggest that there is a low probability that earthquakes occur?*
- *c) Where (geographic distribution) in Greece do a) shallow, b) intermediate depth and c) deepest earthquakes usually occur?*
- *d) Which of the earthquakes, with similar magnitude, typically cause more damages on the land surface a) above the epicentre, and b) at a greater distance from the epicentre—intermediate-depth earthquakes or shallow focus earthquakes? Justify your opinion.*
- *e) List five cities at high risk and five cities at low risk from earthquakes.*

Before the intervention in questions A and E, the students, had on average 40.0% correct answers, while 44.4% answered partially correct and the rest (15.6%) did not answer at all. In question B the majority (84.4%) answered falsely or confusedly and the rest did not give any answer. However, in questions C and D only 3 students were partially correct, 53.3% answered falsely and the rest (40.0%) did not answer at all.

During the intervention with the use of worksheets the students' answers were much improved, not only in terms of the percentages of correct answers but also in a reduction in the percentage of questions that were not answered. Hence during the course using GIS and worksheets that included the above-mentioned questions and some supplementary ones concerning observations of students throughout the process of their work with GIS, all students answered each question. The great majority (80.0%) answered questions A, B, C and E correctly, 11.1% answered partially correctly and only 4 students gave false answers. The only question in which students gave incorrect answers at a percentage of 31.1% was question D, although the majority of the answers were also correct or partially correct.

With regard to the third part of the questionnaire the students answered that they were very satisfied with the outcomes of the lesson and they found it very interesting. The great majority considered that it was relatively easy to finish all the exercises and to continue printing their map, and only five students stated that they had difficulties with the scale of the map and its printing. The final part consisted of open questions, where students were asked to write about their experiences. All of their answers were positive, that it was a fascinating course of geography, very stimulating and they would like having the opportunity more often to participate in such activities.

Discussion and Conclusions

From the results of this research it can be concluded that, with regard to the teachers: a) they were all very positive towards the prospect of GIS use in the class; and b) all of them, regardless of specialty, managed to acquire basic skills in the use of GIS, due to this particular pilot course. Even if most of the teachers initially considered geography and environmental education as the most appropriate subjects for such applications, they in fact revised their opinion after the course, as they found potential use for the exercises and activities in other subjects as well. Furthermore, they stated that if high school conditions improve with regard to pressure of timetable, content of subjects' curricula and infrastructure (modern PC rooms, more work spaces and faster internet access) their use of GIS would be feasible.

Finally, another issue they raised was that of access to digital data concerning different issues related to Greece, Europe or the world. The teachers considered it very time-consuming and difficult if they were required to create their own data sets for a new teaching scenario and to prepare the worksheets and assessment sheets for their students. Therefore, during the evaluation of the second phase of the research, the main obstacle for the teachers that did not implement GIS in their teaching subjects was that they had insufficient time and there was a lack of relevance of the ready-made exercises handed out to them during the course compared with what (subject content) they taught in high school. Moreover, another important obstacle was the availability of PC rooms for the teaching of the relevant courses, as the number of available PCs was small and the students, in order to use them, needed to be separated in two teams, leading to a doubling of the number of hours the PC rooms would need to be used. This problem was also encountered by the Primary and Junior High school teachers, which in combination with that of old PCs, prevented them from running the software in their schools.

Nevertheless, it was extremely important to note that half of the teachers participating in the pilot course made significant efforts to incorporate GIS into their teaching practices in school. However, the most significant success of the programme derived from the participation by the two teachers from the Environmental Education Center, who attempted to create their own data and teaching scenarios and thus produce something highly original and innovative. This ambitious programme will be applied with elementary pupils in the future.

As regards the two teachers who successfully completed the application of the use of ArcGIS with their students, despite the fact that it took place in the laboratory of the geography department of the University of the Ægean, and taking into consideration the availability of existing data and material, they have provided important information, feedback and food for thought and further discussion. The 2nd grade Junior High School students (14 years) demonstrated the competences to utilise GIS effectively and to complete activities, something that was not initially considered possible by their teachers. The enthusiasm and immense satisfaction of the students was extraordinary. Moreover, the degree of accomplishment of the aims of the course was also very important.

In conclusion, the fact that the successful completion of such a GIS application with students who had never used the software before took only two teaching hours, illustrates that, given the right conditions, GIS could be systematically incorporated in their teaching and learning activities. The success of this application was due to the good organisation of the educational material, the analytical instructions for GIS use, the new PCs to run the software, and of course the positive attitude of the teachers towards the programme.

This research therefore demonstrates that, in an ideal situation GIS can be integrated in teaching in several subjects in Greek schools, but is this really feasible in reality? How difficult would it be for the schools' PC laboratories to be upgraded and teacher training courses implemented to disseminate effective methods and approaches? The development of better infrastructure in schools is certainly feasible. Then the only real requirement is the in the hands of education decision makers who have the power to make this type of learning and teaching a reality. Policy makers should therefore seek to find ways to implement GIS in the next Greek curricular reform!

References

Bednarz, S. W. 2004. 'Geographic information systems: A tool to support Geography and environmental education?', *Geo Journal*, 60: 191–199.

Bednarz, S. W. & Van der Schee, J. 2006. 'Europe and the United States: The implementation of geographic information systems in secondary education in two contexts', *Technology, Pedagogy and Education*, 15(2): 191–205.

Demirci, A. 2008. 'Evaluating the Implementation and Effectiveness of GIS-Based Application in Secondary School Geography Lessons'. *American Journal of Applied Sciences* 5 (3): 169-178.

Germanos, D. 2005. 'Issues of spatial organization in creating educational environment for crossthematic curricular', in Germanos, D., Panagiotidou, E., Mpikos, E., Mpotsoglou, K., Birbili, M. (ed.), *A crossthematic approach of teaching and learning in pre primary and early primary school age*, Athens: Ellinika Grammata, pp. 40-51 (in greek).

Houtsonen, L. 2006. 'GIS in the school curriculum: Pedagogical viewpoints', in: Johansson T. (ed), *Geographical Information Systems Applications for Schools – GISAS*, Helsinki: University of Helsinki, pp. 23-29.

Hwang, L. 2006. 'Mapping it out geographic information systems can help administrators make enrolment and facilities decisions', *American school & University*, 4: 34-36.

Johansson, T. 2006. 'GISAS Project in a Nutshell', in: Johansson T. (ed), *Geographical Information Systems Applications for Schools – GISAS*, Helsinki: University of Helsinki, pp. 7-21.

Kerski, J. 2009. *The Implementation and Effectiveness of GIS in Secondary Education: Geographic Information Systems in Education*, VDM Verlag, Germany.

Kimionis, G. 1995. 'GIS as Teaching Tool in Environmental Education: A didactical approach' (in greek), unpublished Msc. Thesis, University of Crete, Rethymnon, p. 120.

Klonari, Aik, & Koutsopoulos, K. 2005. 'Primary and Secondary Educators' Attitudes on School Geography', in: Donert K. and Charzyński P. (eds), *Changing Horizons in Geography Education*, Torun: Herodot Network, pp. 151-155.

Klonari, Aik. 2009. 'GIS in Greek Schools', in: Report of iGuess Project on state of GIS in education, *GIS in schools: State of the Art Report*, pp. 24-25. Accessed May 17, 2013 http://www.iguess.eu/publications.php.

Klonari, Aik. & Laina, V. 2010. 'Primary and Secondary Teachers' Points of View for GIS Use in Education' (in greek), in: *Proceedings of 9th Pan-Hellenic Geographic Conference*, Athens, pp. 796-802.

Kontosi, K. 2007. 'GIS in Education. Possibilities and Perspectives of Use GIS in Teaching Science' (in greek), in: *Proceedings of 2nd Conference on Education*: Language, Thought and Action in Education, Ioannina. Accessed May 17, 2013 http://ipeir.pde.sch.gr/educonf/2/thetikes_epistimes.html

Koutsopoulos, K. 2005. *GIS and Spatial Analysis*, Athens: Papasotiriou.

—. 2010. 'Teaching Geography-Instructing with GIS and about GIS', in: Donert, K. (ed), *Using GeoInformation in European Geography education*, Rome: IGU, SGI, pp. 1-18.

Olsen, A. 2002. 'Using GIS software in school teaching programmes: An initial survey', *New Zealand Journal of Geography*, 113: 17–19.

Patterson, M., Reeve, K., & Page, D. 2003. 'Integrating geographic information systems into the secondary curricula', *Journal of Geography*, 102(6): 275–281.

Rellou, M. & Lambrinos, N. 2008. 'The school Geography curriculum in European Geography education: Similarities and differences in the United Europe', *Pathways in Geography Series*, 36: 1-20.

Pavlopoulos, K. & Galani, A. 2009, *Geology – Geography*, Athens, OEDB.

Rocard, M., P. Csermely, D. Jorde, D. Lenzen, H. Walberg-Henriksson & Hemmo, V. 2007. *Science Education NOW: A Renewed Pedagogy for the Future of Europe*. European Commission, Directorate-General for Research Science, Economy and Society, Belgium, EUR 22846.

Shin, E.K. 2006. 'Using Geographic Information System (GIS) to Improve Fourth Graders' Geographic Content Knowledge and Map Skills', *Journal of Geography*, 105: 109-120.

Sui, Z.D. 1995. 'A Pedagogic Framework to Link GIS to the Intellectual Core of Geography, *Journal of Geography*, 94(6): 578 – 591.

West, B. 2003, 'Students' attitudes and Impact of GIS on Thinking Skills and Motivation', *Journal of Geography*, 102: 267-274.

Yap, Lee Yong, Ivy Tan, Geok Chin, Zhu, Xuan & Wettasinghe, M. C. 2008. 'An assessment of the use of geographical information systems (GIS) in teaching Geography in Singapore schools', *Journal of Geography*, 107(2), 52–60.

CHAPTER ELEVEN

USING MAPS IN DEVELOPING SPATIAL THINKING AND ENHANCE STUDENTS' MATHEMATICAL PROBLEM SOLVING ABILITIES

MARIA PIGAKI

Introduction

Spatial thinking is an intellectual process that is supported by one or more "pictures" of different curriculum subject areas. Consequently, we cannot ignore the fact that developing spatial thinking can be facilitated by the use of various "pictures" of space. As a result, in order to approach and then ascribe space there is a need for such tools as maps recording and depicting it. That is, spatial thinking requires familiarisation in using and approaching space with maps, in a manner similar to language education, which requires familiarisation with the use of writing.

Certainly, the rapid technological developments offer teachers many new opportunities to use digital geographical media in their lessons. However, the ability of students to use technology and treat data, mainly at an early age, is limited in both skill and critical thought. Hence, this gap complicates the learning process and there is thus a need for a preparative technological interdisciplinary tool in order to help pupils to develop major fundamental concepts and skills in different subject areas such as geometry and mathematics. In other words, spatial thinking enables pupils to be critical towards "big data". In addition, given that Geo-media brings the real world into the classroom, a "digital world" approach leads to constructive and active learning practices like problem-solving, project-based learning, fieldwork strategies and an enquiry-based approach.

In addition success in mathematics, among other disciplines, relies on their capacity to approach their subjects spatially. For example, students need to be able to use models, to interpolate and extrapolate, to convert two-dimensional drawings to three-dimensional models and vice versa, to make and interpret graphs and other diagrams and so on. Furthermore, when students use data gathered in investigations for their mathematics courses they encounter many of the anomalies of authentic data: inconsistencies, outliers and errors.

From a dialectical point of view this negates into the important question: how can schoolteachers exploit the advantages of spatial thinking in order to strengthen their teaching of how to solve problems and more specifically to enhance their students' mathematical problem solving abilities.

The answer is simple: using a map, although an absolutely metric tool of space, nevertheless produces an abstract "reality" via simple processes and actions and in this way it helps in the development of spatial thinking, enhancing students' problem-solving abilities.

As a result, it is important to show—and this is the aim of this chapter—that map teaching is an appropriate cognitive tool in developing spatial thinking, which in turn can enhance students' mathematical problem-solving abilities. Within this framework, however, it is necessary to understand mapping as an expression of spatial thinking, which requires a stepwise process leading to the basic components of spatial thinking and its impact on mathematical problem-solving. In order to achieve that, there is a need to examine the role of spatial thinking or how geographic space is perceived.

The teaching experiment was conducted, and continues, in Greek elementary schools (age 9) by permit of the Ministry of Education and the National Pedagogical Institute of Greece. It should be noted that teaching with technology in Greek schools is part of the general aims of the Greek education system; however, its use is more as a teaching instrument and not as a part of the learning process.

The double role of spatial thinking

Perception of geographic space

At the outset it should be stated that the relationship between spatial thinking and mathematics is characterised by a dipole. More specifically, in the learning process of mathematics there exists on one hand the "concrete space", a "dimension" that is created by the experience and the movement of humans (Figure 11.1.), a space in which humans dominate and determine the relations such as neighbourhood, proximity and distance. On the other hand there exists the topologic space, which is a constructive cognitive creation using geometry and mathematics (Figure 11.2.). Thus the term spatial thinking simultaneously describes the real and the cognitive space, and as a tool of teaching space includes these two complementary processes. That is, geographic space is an “abstract” space in its description, and invisible in its relations as opposed to the way it is recorded as a concrete and visible space, which however is an artificial creation using “optical tools” to communicate and transmit information.

This double nature of space allows spatial thinking to become a tool of depiction of geographic space and thus to easily simulate relations such as “distance”, “proximity” and “distribution” that characterise reality. Consequently, the essence of spatial thinking is determined by the interaction of two different cognitive processes:

- The map, as conceptual territorial creation, (space as a subject);
- The map, as “object of knowledge” for the rational comprehension and rationalisation of geographic space (space as object).

Therefore, spatial thinking as a tool of teaching and learning mathematics includes these two complementary cognitive processes that delineate how space is comprehended and are examined bellow.

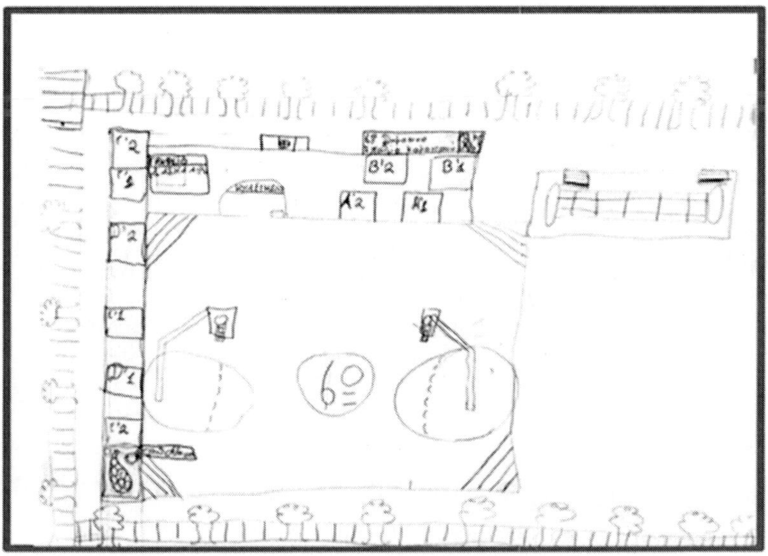

Figure 11.1. Space as a subject.

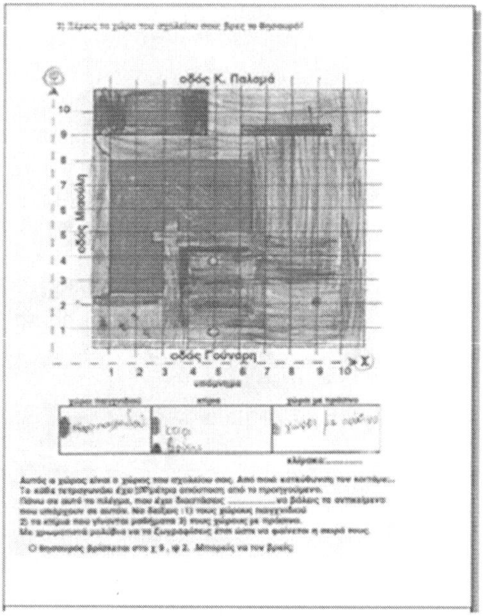

Figure 11.2. Space as an object.

Comprehension of geographic space

For the transmission of spatial thinking that is for space to become perceptible there is a need for a series of processes that in effect constitute the cognitive tool to achieve it. In other words, there exists a need for a consecutive succession of steps, processes based on how space is comprehended and lead from spatial cognition, to scale transformation, to cognitive processes, to mathematical processes, to didactic processes and to mathematical principles as presented in Figure 11.3. This series of steps aims to familiarise children with the three major components of spatial thinking in mathematics, namely topologic attributes, set relations and set theory, which are the result of, as well as determining, spatial thinking. Given, however, the double nature of spatial thinking, there are two different but simultaneous processes that are related to space as a subject and to space as an object.

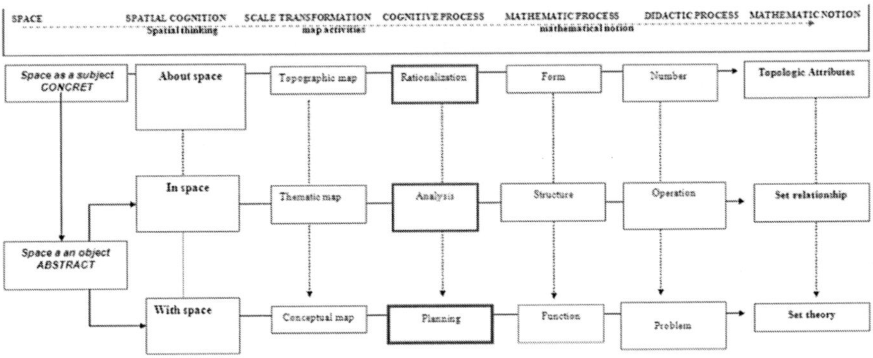

Figure 11.3. Conceptual framework instructing mathematics via maps.

In the case of space as a subject, spatial thinking functions as concrete space and is referred to the transformation of space into cognition about space. This space, with the help of the topographic map, as an instructive spatial tool, through the rationalisation of space promotes students' ability to realise forms and thus understand numbers. Thus this process of cartographic representation develops by induction the skills of understanding topologic attributes, which are necessary in mathematical thinking (Figure 11.3., first line).

In the case of space as an object, where space functions as an object of knowledge, there are two cognitive transformations of space that create the background for spatial thinking. The first of them is related to the passage from abstract space into the cognitive transformation in space. In this space, the thematic map as an instructive spatial tool activates the didactic process of analysis of the characteristics of space and their relationships, which in turn promotes the student's ability to realize structures and thus participate in the cognitive processes of operation. As a result, this cartographic representation develops by succession the important mathematical area of set relationships that composes part of the subject matter of mathematical thinking (Figure 11.3., second line).

The second approach is related to the passage from the abstract space to the cognitive transformation with space, where the conceptual map, as an instructive spatial tool, activates the didactic process of planning, which in turn promotes students' participation in the mathematical process of function and thus their participation in the cognitive process of understanding a problem. This approach of cartographic representation reflects the set theory that constitutes an advance area of mathematical thinking (Figure 11.3., third line).

These processes, although basically constituting an appropriate cognitive tool in developing spatial thinking, additionally represent an instructive framework for teaching and learning mathematics using maps and are presented analytically in the next sections. It should be noted, however, that the above framework was successfully applied in the classroom (9 year-old primary school pupils) and the figures presented are the results of this application.

Space as a subject

The first of the three distinct approaches to space, which is associated with a host of processes described below, is the one that regards space as a subject.

Spatial cognition: about space

In order for a map to become a tool in enhancing spatial thinking there must be a passage from the concrete subjective space to the space-creation. That is, children have to change their perceptions into reality. In other words, they need to learn when seeing "pictures" of geographic space to change them into geometric forms, which include 2D and 3D geometrical spatial entities. In this way, spatial thinking functions as an instrument of transforming space through a series of spatial decisions and processes which in turn involves different levels of actions leading to the cognition about space.

Scale transformation: topographic map

The passage from subjective space to "real" or Euclidian space is achieved through an instructive process that is related to the geometric expression of the elements of space (mapping). That is, the description syntax and scale of the topologic elements is important knowledge, for they represent and correspond with the reality of geographic space. As a result, this process is, in essence, aiming at changing students' overall view, as well as transforming the forms of space, so that students can be helped to "see" the abstract space. Consequently, the Euclidian space provides all the essential tools for the process of this cognitive transformation, which with the help of topologic attributes constitutes the base for incorporating spatial thinking.

More specifically, the recording of a familiar element of space in terms of absolute location, which is provided by the syntax of a topologic attribute, facilitates space recognition by providing understandable forms and shapes that simplify it (Figure 11.4.). But through these forms and shapes, space is progressively revealed by continuously providing more complex information, which in turn leads to the comprehension of relative location in space. Furthermore, the scale, in absolute terms, is the first step in observing and reading space and leads into a second stage: the scale of relationships 1:n. That is, students can be introduced into the significance of relationships.

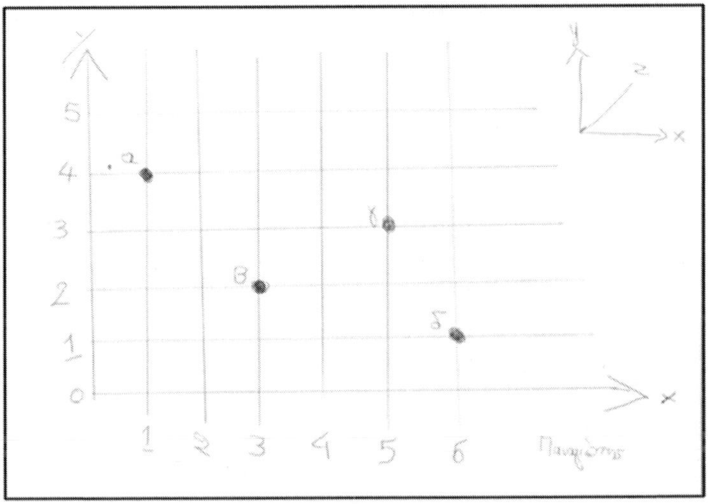

Figure 11.4. Absolute location.

Didactic process: rationalisation

Children should be in a position to recognise locations, to identify in a "vertical" reading the forms of geographic space and mainly to record as forms or shapes the geographic entities. These processes together with the notion of scale contribute to students' didactic growth and constitute the base of rationalisation of space. Consequently, the familiarisation of students with geographic elements requires the convergence and the activation of the processes that are provided by the use of topologic attributes as tools of rationalisation of space.

Mathematical process: form

The mathematical development of form includes processes that support reasoning and calculating. That is, children visualise: relations, transformations from 3D to 2D (Figure 11.5.a), rotations of objects to view their other sides (Figure 11.5.b), creations of new viewing angles or perspectives and remembering images in place and space. In addition, an object can be specified in relation to the observer, to the environment, to its own intrinsic structure, or to other entities in real space. That is, each instance requires the adoption of specific spatial frames of reference or context. (Figure 11.5.c). In other words, spatial thinking begins with the ability to use space as a framework.

Figure 11.5. Rotation (a), From 3D to 2D (b), Form (c).

How children view transformation leads them to define the form of spatial elements, to think about "distance" in space, to attempt methods in order to place it in space. In fact, it helps to recognise forms, to constitute notions such as: far, near, into, content and to use typical operations to reconstitute the holistic view of space (Figure 11.6.).

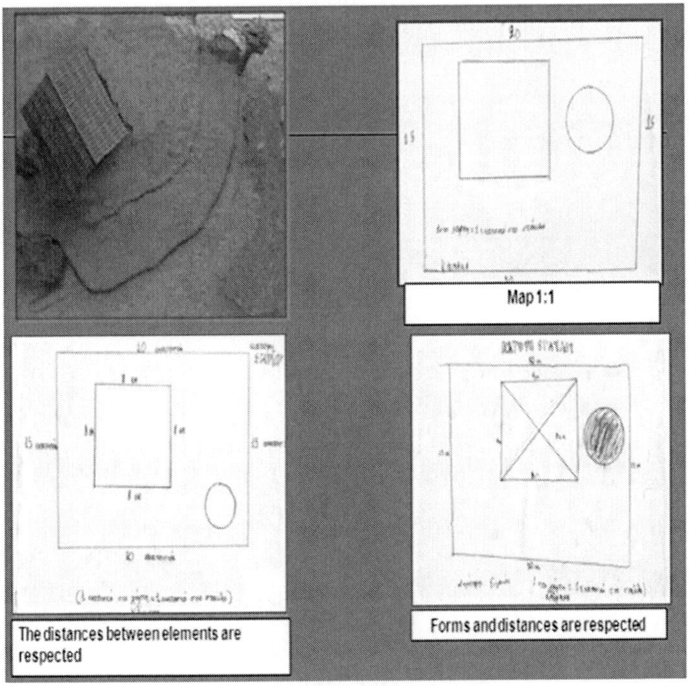

Figure 11.6. Distance, forms and notions.

Cognitive process: numbers

Students' perceptions of space by using cognitive "abstractions" constitutes the beginning of its comprehension. More specifically, with the use of a typical operation (number) to record space, which is accomplished by using proportions and geometry, students see and conceive location in space. At the same time, they simultaneously shape the "words" that help them understand the complex meanings of the spatial language, which lead them into comprehending the expressions of mathematical principals. Moreover, through the Euclidean recording of space students observe and recognise the dynamic and complex nature of space. That is, through the process of calculating they conceive the existence of many parameters behind the spatial phenomena (Figure 11.7.).

As a result, by understanding the relationship between numbers and space children rationalise the mathematic depiction of space. Given that the typical operations used by children are: addition 1+1+1+1=4, division

4=1+1+1+1 and multiplication 1x4=4, by making use of numbers children construct the forms of the spatial elements. For example: $E=a+\beta+\gamma+\delta$ (Area A = Area α+ Area β+ Area γ+ Area δ). (Figure 11.7.)

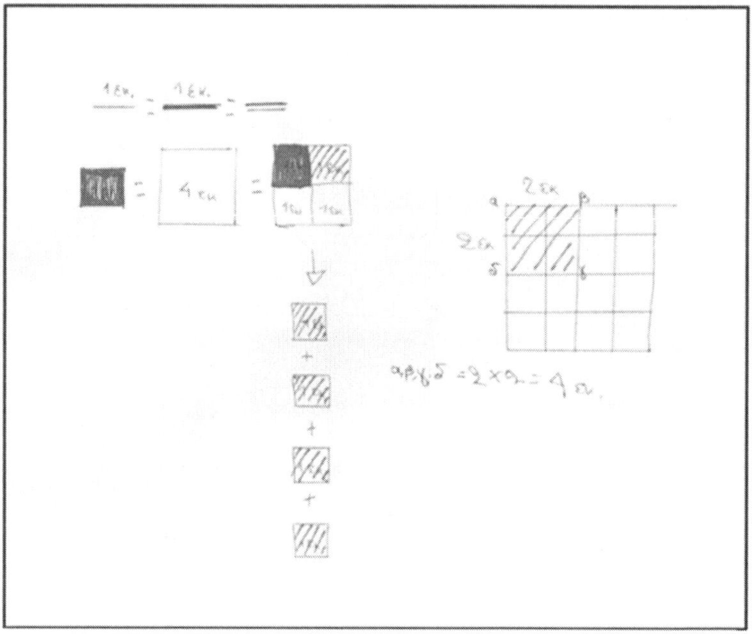

Figure 11.7. Proportion.

Mathematical principal: topologic attributes

The recording of an element of space familiar to students and related to absolute location is provided by the syntax of its topologic attributes. More specifically, through forms and shapes, space is progressively revealed by providing continuously more complex information, which in turn leads into the comprehension of absolute location in space. As a result, scale, in absolute terms, is the first step in observing and reading of space and leads in a second stage to a scale of relation 1:n. That is, students are introduced to the significance of topologic attributes. In other words, by facilitating spatial recognition in understandable forms, locations and shapes this tool simplifies it. Consequently, the initiation to skills such as the transposition from a partial approach to space to an integrated one reveals elements of space that constitute part of spatial thinking. Thus, children understand the qualitative properties of spatial elements that are invariant under certain kind of transformations. In addition, they recognise spatial elements as a point (no area), a line (a set of vectors) or a polygon (an area) as well as comprehending notions such as convergence, connections and continuity. In this way children can describe space in mathematical symbols: A=α', Γ=αβ+βγ+γδ+δε+εα (Figure 11.8.).

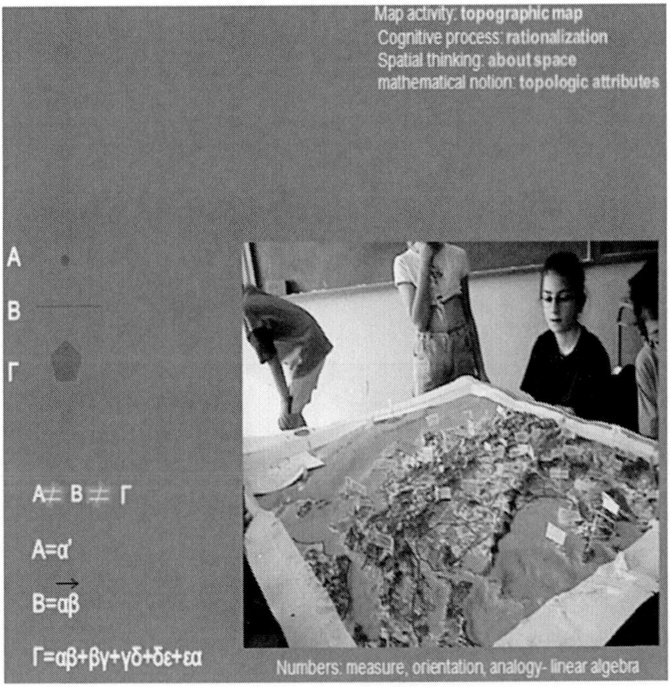

Figure 11.8. Topological attributes.

In sum, the spatial thinking approach to space via topographic maps leads children to understand the form of spatial elements that is created by numbers and measures, and finally realise space as a set of relationships.

Space as an object: in space

The case of space as an object, that is the transformation of the space-object into a tool of spatial thinking, is related to two processes: the first refers to the operation of spatial relationships (cognitive transformation in space leading through a set of processes to set relationships), and the second to the creation of spatial systems (cognitive transformation with space leading through the same set of processes to set theory). Next, the set of processes related to the first transformation, which refers to "in space", are described.

Spatial cognition: in space

The spatial cognition in space, available to children through the use of symbols (colours, patterns, size, etc.), helps them to draw inferences and analyze relationships, changes, patterns and processes taking place in space as well as familiarizing them with the spatial system.

Scale transformation: thematic map

Variables expressing spatial relationships greatly help students to comprehend such principles as similarity, diversity, hierarchy and quantity and thus reveal progressively the "invisible" space of topologic attributes. This, of course, leads into projecting relationships in space as a unified geographic framework. As result, the introduction to the principles, the rules and the syntactic of spatial representation initiate cognitive processes that lead students towards spatial thinking. Consequently, the composition of a map, and specifically a thematic one, is an essential tool in providing students with dexterities to treat and generalise spatial elements (Figure 11.9.a). More specifically, the transposition via thematic maps helps students comprehend the internal relationships of these elements, discover phenomena in space, analyze structures and finally realise space as a set of spatial elements.

Cognitive process: analysis

The process of analysis is expressed by at least three questions: "Why is it there?", "How did it get there?", and "What is its significance?" In terms of spatial thinking therefore, the focus is on how students can attribute meaning to what they observe and mainly to explain it. That, of course, involves the teaching of the analytical concepts and principles that provide them with the ability to comprehend, to see connections and to explain patterns and processes in space. But these abilities are easily provided through the creation of thematic maps, which help to comprehend relations and see connections between different entities in space.

Mathematical process: structure

Structure is a fundamental, tangible or intangible, principle referring to the recognition and observation as well as to the nature and permanence of patterns and relationships of entities. Based on this, students using symbols as an instructive spatial tool can activate the process of realizing structure, which in turn promotes their participation in mathematical problem-solving. In addition, there are systemic tools (i.e. a set of operations) available to students, which can help them in drawing inferences and in analysing relationships, changes, patterns and processes taking place in space. Students can therefore use such tools successively to participate in the cognitive process of mathematical structure, which in turn lead to the cognition in space.

More specifically, children recognize that a structure is a set of spatial elements consisting of mathematic entities, and can visualise them with symbols and inject them with meaning or significance. Such structures are: topology, assimilation, differentiation, hierarchy, etc. Sometimes, a set of spatial elements is endowed simultaneously with more than one structure offering mathematicians a more in-depth study. For example, if a set has a structure, it creates a group whose members are related in a certain way. That is, because of their structure, the set becomes a new topological group where: A>B and A<a’ or B<a’ (Figure 11.9.b).

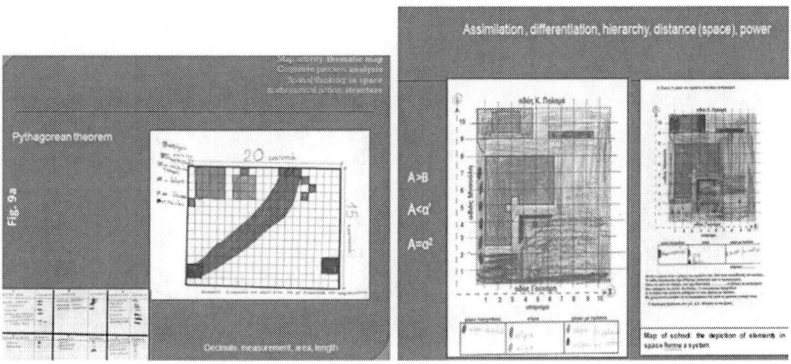

Figure 11.9. Space as a set (a). Space as a structure (b).

Didactic process: operation

Operation is an action or procedure, which produces a new value from one or more input values. Such action includes addition, subtraction, multiplication, and division. Those procedures, however, additionally imply the notion of "distance", which contains two expressions: the "distance space" and the "distance time". As a result, given that distance is a fundamental component of cartography, such operations help students to analyse spatial elements in order to structure the spatial relationships in space and to depict the changes over time. (Figure 11.10.a and Figure 11.10.b).

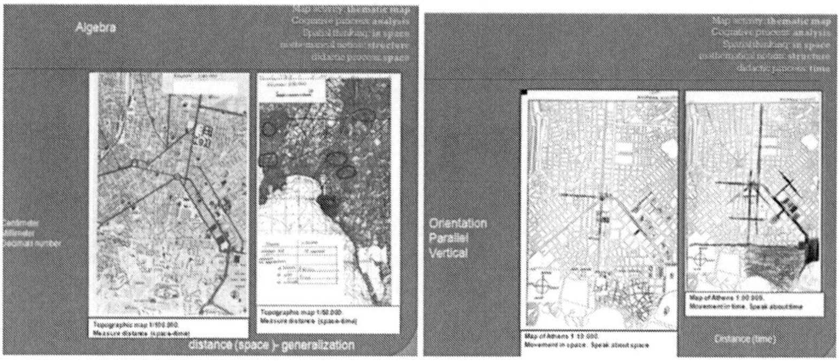

Figure 11.10. Distance space (a) and distance time (b).

Mathematical principal: spatial relationships

A representation tool (i.e. a thematic map) using an optical "dialect" transforms the elements of geographic space into shapes or forms and expresses spatial phenomena using aspects of mathematical principles. In this way, a spatial relationship can be represented by a map and this representation can in turn be an essential tool in providing students with dexterities to treat and generalise spatial elements. As a result of this transposition students comprehend the internal relationships of these elements, discover phenomena in space and become observers of their development. In addition, the description of structure implicitly offers an account of what a system is made of, such as configuration of items or a collection of inter-related components. A structure may be a hierarchy (a cascade of one-to-many relationships), a network featuring many-to-many links, or a lattice featuring connections between components that are neighbours in space. (Figure 11.11.)

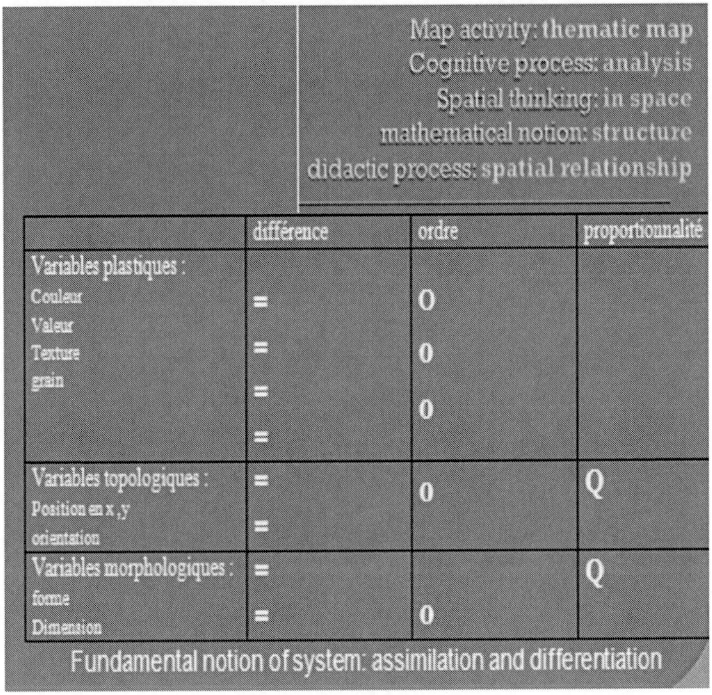

Figure 11.11. Notional system.

Variables expressing spatial relationships help in comprehending principles such as “similarity”, “diversity”, and “hierarchy”, and thus reveal progressively the notion of a system, or, in other words, the spatial organization. This, of course, leads to projecting relationships in space as a unified geographic framework. As a result, an introduction to the rules and to the syntax of spatial depiction activates didactic processes that refer to the triplet: form, structure and function and they constitute an optimal tool in strengthening spatial thinking in order to enhance the student’s mathematical problem-solving abilities.

Children apprehend space via the application of cartographic symbols. They realise the problem in space as a mathematical relation in such a way that each spatial element of a given set of elements (the domain of the function) is associated with another spatial element of another set of elements (the range of the function). As a result, they understand and utilise principals such as:

- Transitivity—(logic and mathematics) is a relationship between three elements in such a way that if it holds between the first and second, it also holds between the second and third and thus it must necessarily hold between the first and third; and
- Reflexiveness, reflexivity—(logic and mathematics) is a relationship that holds between an element and the type of relationship it contains.

Space as an object: with space

The second case of space as an object, where space functions as an object of knowledge, the cognitive transformations of space that create the background for spatial thinking are related to the passage from abstract space into the cognitive transformation with space. That is, it refers to the representation of maths models, which basically express the set theory, and is examined next.

Spatial cognition: with space

The passage from "in space" to "with space" refers to a cognitive transformation where the map constitutes a spatial model, representing a tool to contemplate and analyse and which helps to describe relations and overlays that at the end might lead into restructuring space itself. The definition of space in terms of points, lines and polygons is not simply a way of recording places or their characteristics, but indicates a cognitive model of space that describes such principles as convergence and divergence, thus promoting spatial thinking which is absolutely necessary in teaching and learning mathematics.

Scale transformation: mathematical models

Mathematical models render space itself as a tool of contemplation and research, which contributes to problem-solving and thus involves the utilisation of spatial thinking. As a result, the objective of this spatial representation approach is to promote knowledge itself. That is, there is no interest in attaining new spatial knowledge, but rather to comprehend and apply existing knowledge. These, in turn, help students to develop an organised and critical spatial perspective that promotes problem solutions, one of the basic elements of mathematical thinking. Therefore, since the specific spatial representation depicts the internal structure of space, students understand the application of those systems imposed by human activities and the natural environment. Undoubtedly, the use of mathematical models is not based on the typical scheme: locate, analyze and construct, but rather on the genuinely conceptual procedures of: analyzing-inventing-applying. These, in turn, help students develop an organised and critical spatial perspective and try to resolve problems related to real life. This procedure constitutes the third of the basic elements of spatial thinking, with space. (Figure 11.12.)

Figure 11.12. Conceptual procedures of analyzing-inventing-applying.

Cognitive process: planning

The cognitive process of planning involves forming problem-solving models, formulating solution to problems and in general applying spatial thinking in a practical manner to real life situations that exist around us. As a result, children have to comprehend the contents of space, which can efficiently be provided by the appropriate use of models. Thus mathematical models are important tools in spatial planning, by facilitating the development of the cognition with space.

Mathematical process: function

This process involves the integration of spatially referenced data in a problem-solving environment. That is, mathematical models constitute tools for facilitating spatial thinking and not an end themselves. They are used in formulating hypothesis and in confirming the dynamic nature of space. They allow students to devise graphic subterfuges in order to apply spatial thinking in examining the relationships of people, places, environments and thus developing intellectual dexterities that allow them to move from absolute to relative and functional space. Consequently mathematical models shape and strengthen spatial thinking.

The symbolic language paradigm of mathematics takes the concept of variables and functions to a new level. In mathematics a variable not only stands for a value, but can also be used purely symbolically. Thus building on mathematics' powerful pattern language, "functions" can be defined not just as instruments of arguments, but also as a device to transform a pattern with any structure. Children study the rules of relational operations that are the constructions and concepts arising from the symbols including geometry, analysis and topology and which introduce notions such as groups and rings, which are all studied in the area of mathematics called abstract algebra (Figure 11.13.).

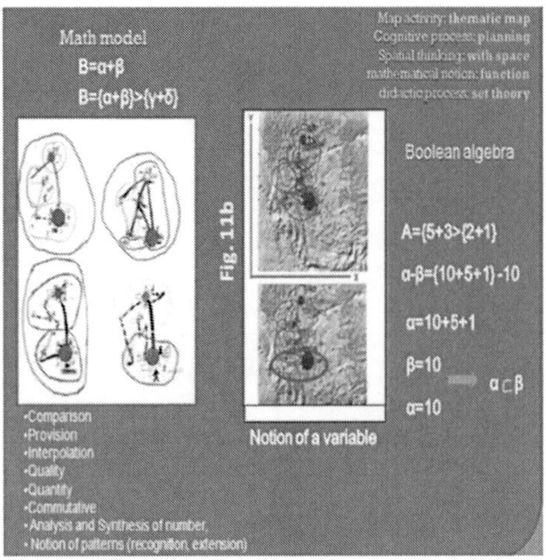

Figure 11.13. Mathematical power.

Didactic process: problem

The didactic process of problem-solving helps pupils to: identify changes in the uses of, and regional differentiation of, space(s); measure the physical arrangement and clustering of phenomena to detect spatial patterns; document spatial patterns over time to infer processes; study flows (e.g. migration, trade, and shopping patterns) between specific locations as indicators of spatial-temporal interactions; and measure spatial associations (and space–time associations) for testing hypotheses. (Janelle, D.G. and Goodchild, M.F. 2009). In sum, spatial questions involving the ability to use the results of planning to solve problems or form judgments about a problem, simply express the process of set theory.

Mathematical principal: set theory

The passage from the space-object to the cognitive transformation with space refers to a cognitive transformation where spatial models, representing a tool of contemplation and analysis, help to describe relations and overlays, which in the end might lead to restructuring space itself. The definition of space in terms of points, lines and polygons is not simply a recording of places or their characteristics, but promotes a cognitive model of space that describes such principles as convergence and divergence and thus promotes the notion of algebra (Figure 11.14.).

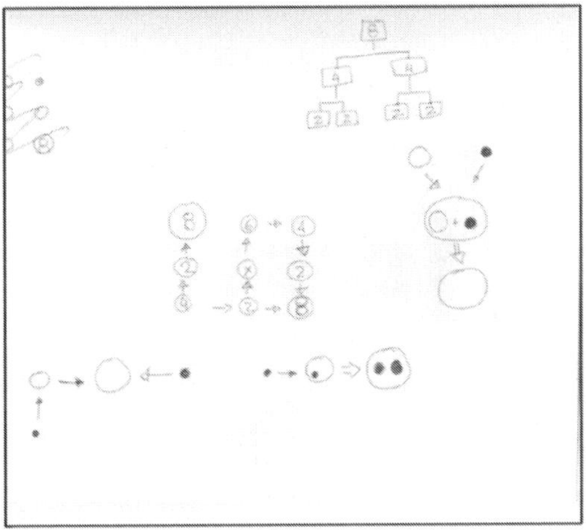

Figure 11.14. Notion of algebra.

Set theory begins with the fundamental relationship between an object and a set. Since sets are objects, the membership relation can equally relate to sets themselves. If all the members of set A are also members of set B, then A is a subset of B, denoted A $\subseteq$ B. From this definition, it is clear that children can conceive space by collecting activities (therefore constructing sets=planning), giving them the ability to apprehend mathematical notions using a given problem and thus plan with space.

, By understanding and using among others the following operations children can therefore approach and then ascribe space:

- Union of the sets A and B, denoted A $\cup$B, is the set of all objects that are a member of A, or B, or both. The union of {1, 2, 3} and {2, 3, 4} is the set {1, 2, 3, 4};
- Intersection of the sets A and B, denoted A $\cap$ B, is the set of all objects that are members of both A and B. The intersection of {1, 2, 3} and {2, 3, 4} is the set {2, 3}.

Conclusions

Teaching in elementary school must shift from "teaching with data" to "teaching with technology" resulting in what is now termed "datafication", which is a process that takes into account the vast possibilities offered by the "digital world", including Google, social networks, GIS etc. In other words, for "teachers to bridge the technological gap between students and themselves, they need to use geo-media in the classroom" (Digital Earth 2013). As a result, one of the most important objectives regarding spatial thinking, that of improving students' ability to manipulate the overabundance of information, can be achieved as part of their digital literacy. Cartographic representation therefore becomes a preparative interactive time-space tool, which enhances students' perception in such a way that they can understand the conflicting reality, resulting from the interaction of time, space, and man. That is, in order to understand the ever-changing spatial reality, students are trained in logical procedures that enhance mathematical thought. Students therefore perceive reality as a continually changing entity, where there are many factors playing an important role and where all the ingredients that represent spatial thinking are interrelated.

Spatial thinking may function as a vehicle for a deeper understanding of mathematics and for the integration of mathematics and geosciences, and it thus represents a fundamental form of inquiry. As a result, the use of maps in developing spatial thinking may result in opportunities to advance education in mathematics beyond purely descriptive studies and to provide mathematics classrooms with authentic cases of problem-solving. Spatial thinking thus may function as a link across the curriculum and may help the curriculum to become coherent.

References

Koutsopoulos, K., Pigaki, M. and Dimitrov, S. 2008. "Examples of Teaching Geography Using GIS", in *HERODOT Workshop: Teaching Geography with GIS*, Conference proceedings, Albena, Bulgaria, 13-16 June 2008.

Leininger-Frézal, C. & Pigaki, M. 2013, "From teaching Geography of information towards Geography of location and attributes" in *EUROGEO Conference proceedings*, Bruges.

Pigaki, M. 2011, *Enseigner la Cartographie: la Carte, un outil rationnel pour la compréhension spatial. Expérimentation pour une élaboration des programmes scolaires,* Editions Universitaires Européennes (EUE), Sarrebruck.

CHAPTER TWELVE

LEARNING GEOGRAPHY AND GEO-MEDIA

MARIA LUISA DE LÁZARO,
MARÍA JESÚS GONZÁLEZ
AND MARÍA DEL CARMEN MÍNGUEZ

Introduction

This chapter is based on a Master's level geography Project for the training of future teachers (2011-2012) at Complutense University and the University of León. The objective of this type of project is to improve teaching quality. Digital literacy related to geo-media and geographical concepts (spatial thinking) are promoted for future teachers. In order to achieve this, it is necessary either to create new teaching materials or reuse open data in an educational way. Web 2.0 has been very useful towards fulfilling this need. Part of this work was collected and put on a web page that can be visited via the Complutense University Moodle virtual campus.

Web 2.0 has really changed our daily lives and the ways of communication and collaboration. This, along with ICT infrastructure improvements, gives us unprecedented chances to change the methods used to study geography. The main aim of our work was to take advantage of ICT, geo-media, Web 2.0 and the geo-information resources on the web (González and Lázaro, 2012) in order to achieve ICT competences with young learners and in the pre-service training of future geography teachers in the way that the European Digital Agenda indicates (Lázaro, Alcolea and González, 2012).

Objectives

The main aim of the project was to create teaching materials for learning the geography of Spain using geo-media and Geography Information Technologies (GIT). It reinforces spatial thinking focus on geo-referencing and geo-localisation in relation to space or land 'territory'. Teachers in future can use some of the updated practical examples.

The materials and resources have been linked to a webpage created on the Moodle virtual campus of the Complutense University (Lázaro, 2012). With them we try to:

- Explain in greater depth, the main geographical concepts illustrated by examples throughout Spain;
- Design a working methodology based on observation, review and assessment of landscapes through photographs and videos;
- Reinforce the observation of landscape and spatial competences of students through geo-media and GIT tools (whiteboard, virtual campus Moodle, GPS ...);
- Design quality and original materials to create virtual campus activities and take advantage of communications through Web 2.0;
- Strengthen the dynamics of group work, including the debate, creation, selection and implementation of materials developed and their projection in university lectures by encouraging good teaching practices;
- Create a webpage with the developed materials so that teachers and students who have participated in the project can access them as well as any other member of the teaching community.

Methodology

Active participation by all the team had been very important. The work has developed in three parallel directions to have the necessary feedback for carrying out the project: 1. Research and workshops by the team, as a main source of advancement. This arrangement/preparation consists of the identification, selection and own creation of the necessary educational materials essential for its later use by the team; 2. Training aspects in using GPS and other geo-media to start the project and improve the geo-media competences of students—future teachers; 3. Apply geo-media knowledge and skills to lessons and fieldwork. The University virtual campus Moodle was considered a good tool for sharing teaching materials.

The work was based on landscape study and geo-media image commentary (either photographs or videos) drawn up by the team. A map of Spain was created in ArcGIS Online with the images located on the map. A professional video about particular landscapes such as *dehesa* (http://youtu.be/tJ_qSCrr2_E), limestone or granite was also created. Dissemination of the project has since been a strong component.

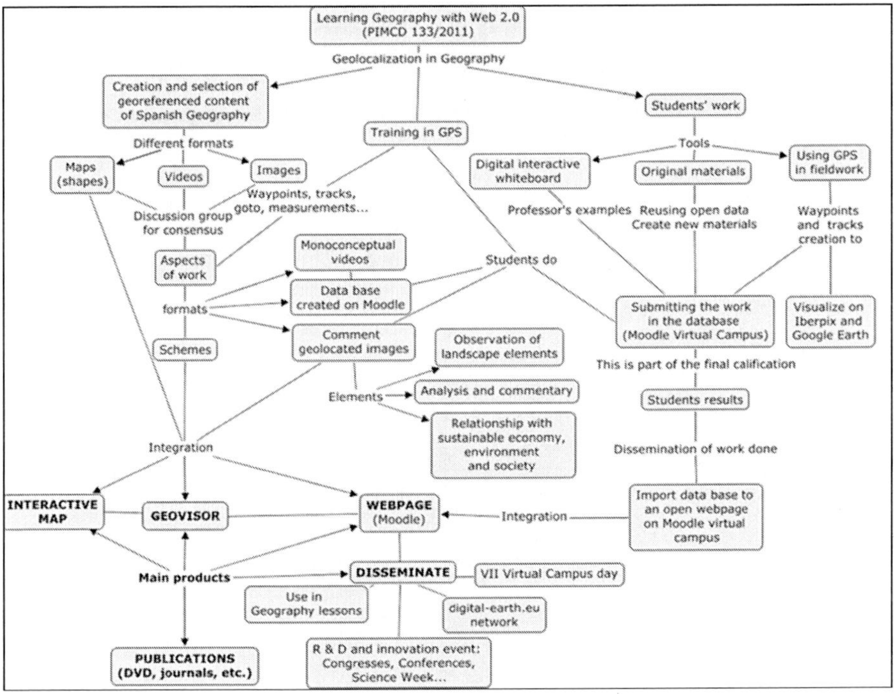

Figure 12.1. Steps of the work.

Results and materials

The approach to learning about the diverse landscapes of Spain has been carried out from a sustainability perspective (Minguez, 2010) as a medium to achieve geographical competences. Text comments, data, graphs, maps and so have been very helpful for the students. The team has created many different types of geo-media materials. Some examples are illustrated here.

A hot map with photographic comments

A hot map was put together by the Project team with items to review and analyse located in the different provinces (Figure 12.2.). Clicking any marked point on the image opens a pdf with geographical data of the image. Many of these images had been taken during fieldwork and are used by students to learn how to use a GPS and obtain coordinates.

Fieldwork took place in several different types of landscape such as Sigüenza (Guadalajara), Toledo, Manzanares El Real and Paracuellos de Jarama (Madrid), Talavera de la Reina (Toledo), Iruelas (El Tiemblo, Ávila), León, etc. Students also collected their own photographs to include with their comments in the Moodle database as part of their work on the subject.

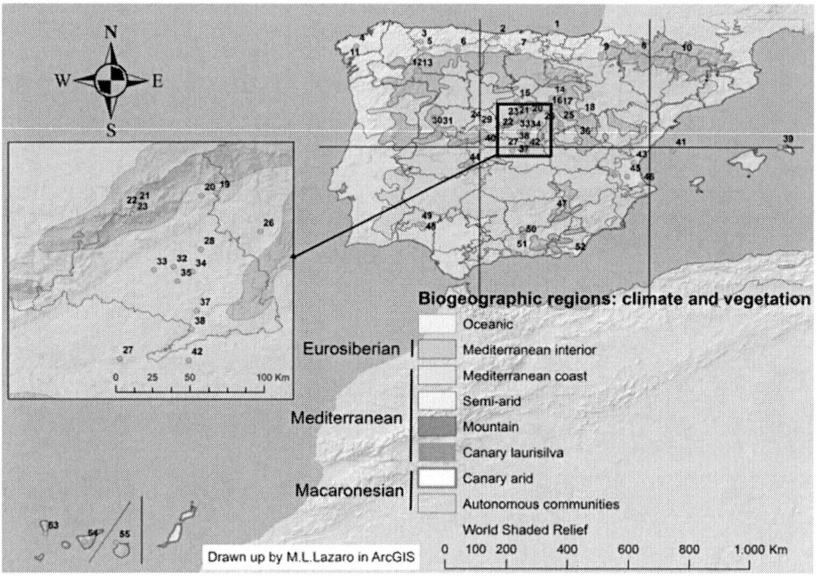

Figure 12.2. Locations of commented images.

Comments were made on photograph images following a common sketch template created by the team of students (see Annex I) beginning with a significant image title including the subject, municipality, province, and autonomous community. The image comments begin with a brief description of landscape elements (observation of landscape elements) and an explanation in the analysis and comment. At the end, the students broadly describe the image in its context in three sustainable dimensions: environmental, economic, and social. An Internet link and a key book or article is also added to the information as well as the location of the image, date, course, altitude and exact geographical coordinates where the image was taken. Some key words and the coordinates can help teachers to locate the main subjects of the image and to look for additional information on either web 2.0 (virtual globes, viewers…) or in books.

Videos

Taking into consideration that it is easier to learn some concepts with videos or moving images, the team worked with students in several ways, either to make their own video, or how to look for thematic videos and download them from the Web. Videos were made by the team on granite, limestone and metamorphic landscapes and are made available on YouTube.[1]

In examining sustainability other detailed information was also collected. For example, of different fishing techniques (in Spanish)[2] such as trawling, purse seine fishing, *almadraba*, long line craft and other techniques, which are very harmful to the marine environment, such as trammelling, now illegal, and driftnets prohibited on the high seas by the EU in 1992 (http://www.youtube.com/watch?v=PYPMjpPlWqg, uploaded by OCEANA: Protect the oceans). Fishing, often using fishing rods, or other systems can sometimes occur without death, i.e. returning the fish taken from the sea, for example *Aquaturismo*: http://www.youtube.com/user/txoacuariooceana (uploaded by OCEANA).

It is possible to view the wide variety of existing fishing techniques on YouTube, their adaptations to environmental balance, their relationship with the depletion of fish stocks, and their associated problems with humans. This approach encourages a reflection on the sustainability of activities at sea and the importance of good governance.

---

[1] http://www.youtube.com/watch?v=xx5wvOEz5TE (granite landscape)
http://www.youtube.com/watch?v=hBUAyPzr88s (limestone landscape);
http://www.youtube.com/watch?v=29ygmwLWsWc (metamorphic landscape)

[2] Trawling: http://www.youtube.com/watch?v=srUhwGk6Cok involves a network that is sweeping the seabed movement regardless of fish stocks. It is considered very harmful to the marine environment; Purse seine fishing: http://www.youtube.com/watch?v=XViuKk8x8bw, a network that entails a school of fish, sometimes previously detected by radar. Typically used for catching fish that usually travel in large schools, such as sardines, mackerel...; *Almadraba*, requires installation of a maze of nets for fishing tuna in their migration process. The example of the trap of Conil, Zahara (Cádiz) shows the traditional work in the trap: http://www.youtube.com/watch?v=ZWB9HBGXYv8. The *Almadraba* of Barbate (Cadiz) shows us the work in a trap modernized: http://www.youtube.com/watch?v=pc63pWAMrAY; Long line craft: http://www.youtube.com/watch?v=W4j5lc1tLOI is a large net with a type of fish hook at the bottom. It is used to capture species existing in the greater depths of the seas, hence its name.

Database

A cooperative database on Moodle allowed the sharing of materials between students. It also offered the chance to integrate materials from other courses. The resultant web page was open access and based on a URL that can be seen by anybody on the Internet. The database contained a number of fields filled and uploaded by the students (Figure 12.3.). This database allows searches to be done according to type of material (video, graph, map, image, GPS data, etc.) and the area and scale it relates to, namely global (world), regional (Europe, Spain…), local (Madrid…).

Figure 12.3. Data base created on Moodle.

The proposed objective at the beginning of the database work was to share materials between students and teachers. The context applied was that it should be original or open public data and be connected to spatial thinking (Zwartjes, 2012). The items on the database should also take into account geographical science procedures that are the basis for the acquisition of skills (e.g. comments in landscapes through fieldwork photographs, maps, graphs, tables, texts, statistics, etc.). The teachers presented and explained the tasks in the classroom and described procedures giving some examples.

The work resulted in 135 records (two per student). 64 records (some were photographs) were elaborated or used by students (40%); 53 were taken from open or public data (statistics mainly) (35%). The rest were unopened resources and were removed from the database (15%). We can say that students did some remarkable work, i.e. when seeking and selecting the necessary public information. The systematic review of different resources also improved the digital competences of the students and staff members. All of this information can be found on the Web at https://cv2.sim.ucm.es/moodle/course/view.php?id=12502&lang=es_utf8. It has also been produced on a DVD, organised into two main parts: the interactive (hot) map, and all the information from the project, the video and the web pages (Figure 12.4.)

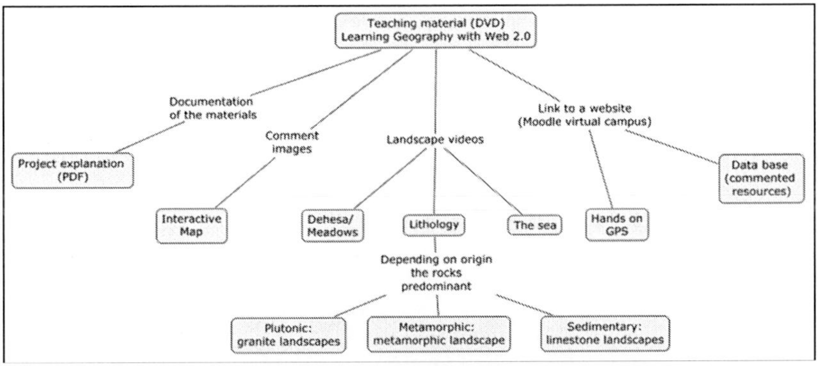

Figure 12.4. Sketch of the webpage and/or DVD.

The webpage provides the database and some advice on procedures and some teaching examples such as the comments (mentioned above) on an unemployment interactive graph produced on the Google Public Data Explorer with Eurostat data. The same graph can be seen in English and other languages.

All of these initiatives have increased interactive work and group work during university lectures amongst students and professors.

ArcGIS Online

After the project work had been collected, the decision was taken to use ArcGIS Online with the same purpose as the hot map and database activity. All the information was filled in on an Excel™ sheet (Figure 12.5.).

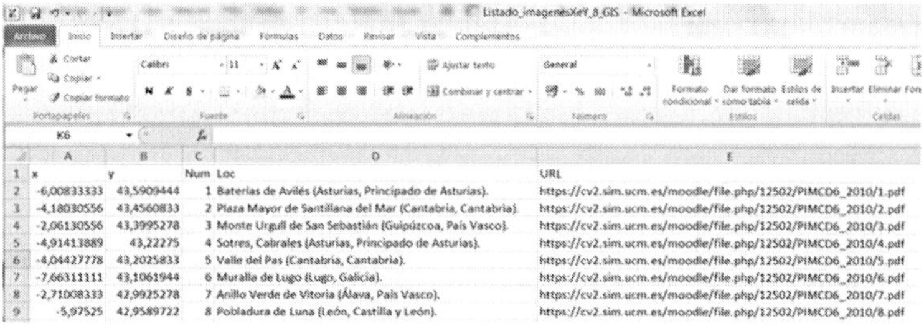

x	y	Num	Loc	URL
-6,00833333	43,5909444	1	Baterías de Avilés (Asturias, Principado de Asturias).	https://cv2.sim.ucm.es/moodle/file.php/12502/PIMCD6_2010/1.pdf
-4,18030556	43,4560833	2	Plaza Mayor de Santillana del Mar (Cantabria, Cantabria).	https://cv2.sim.ucm.es/moodle/file.php/12502/PIMCD6_2010/2.pdf
-2,06130556	43,3995278	3	Monte Urgull de San Sebastián (Guipúzcoa, País Vasco).	https://cv2.sim.ucm.es/moodle/file.php/12502/PIMCD6_2010/3.pdf
-4,91413889	43,22275	4	Sotres, Cabrales (Asturias, Principado de Asturias).	https://cv2.sim.ucm.es/moodle/file.php/12502/PIMCD6_2010/4.pdf
-4,04427778	43,2025833	5	Valle del Pas (Cantabria, Cantabria).	https://cv2.sim.ucm.es/moodle/file.php/12502/PIMCD6_2010/5.pdf
-7,66311111	43,1061944	6	Muralla de Lugo (Lugo, Galicia).	https://cv2.sim.ucm.es/moodle/file.php/12502/PIMCD6_2010/6.pdf
-2,71008333	42,9925278	7	Anillo Verde de Vitoria (Álava, País Vasco).	https://cv2.sim.ucm.es/moodle/file.php/12502/PIMCD6_2010/7.pdf
-5,97525	42,9589722	8	Pobladura de Luna (León, Castilla y León).	https://cv2.sim.ucm.es/moodle/file.php/12502/PIMCD6_2010/8.pdf

Figure 12.5. Excel sheet for making the map.

The map was created using the spreadsheet information. The astounding results (Figure 12.6.) led us to consider its potential. On the map, the hyperlink "*Más información*" opens the commented landscape sheet, which contains a photograph.

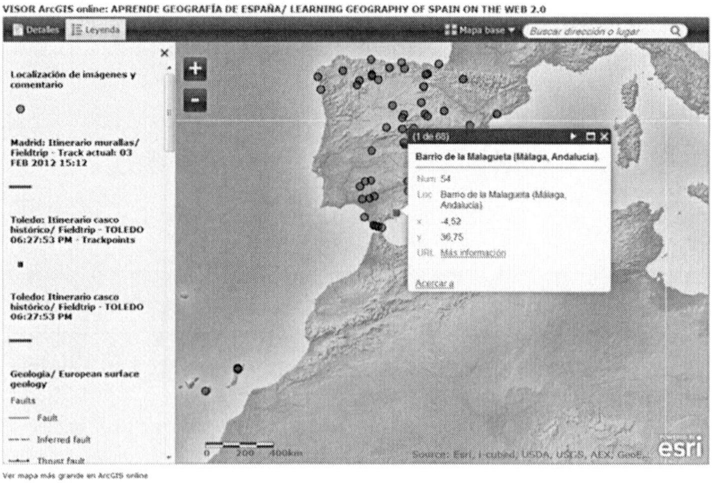

Figure 12.6. Our map on ArcGIS online.

The same work was therefore done easily and quickly in a Cloud-based system, and it wasn't difficult to cooperate and to share materials. The analysis of one's own information, in this case the photos and landscape comments and some GPS routes, could readily be enhanced by finding other geographical information on maps from many organisations e.g. ESRI (base map, satellite images…), the EU (geological, Corine…), or the National Geographical Institute, for instance. ArcGIS Online thus offered the possibility to integrate open, public, geographical data.

Conclusions

This case study shows that it is possible to enhance the learning and teaching of geography through the available GIS and Web 2.0 tools. The experience has motivated the group. The database has been greatly enhanced by a using a Cloud-based GIS tool to create a resource on Spanish geography. It also integrates other GIS tools in geography lessons (Lázaro and Gonzalez, 2005) as other countries do (Milson, 2011; Demirci, Milson and Kerski, 2012). The use of ArcGIS Online has been very helpful and has proved that geography learning can be supported through GIS tools.

The project has improved the awareness and use of geo-media (Buzo, 2012), ICT competences, and spatial thinking with the initial training of future teachers in teachers' lessons for learning geography. The Moodle virtual campus and Web 2.0 have offered very useful tools. Teamwork and collaboration were very helpful in improving ways to teach geography. The interactive work between teachers and students, both together as well as on their own on national and international projects, like digital-earth.eu, has been helpful in expanding the content and ideas of national and local projects. It also allows researchers to convey their experience and research results to other teachers as it does to the society.

References

Buzo, I. 2012. Utilización de Geomedias en la Enseñanzas de la Geografía. [on-line] *Comunicación al Congreso Nacional de Contenidos Educativos Digitales*, Mérida.

González, M. J. & Lázaro, M. L., La geoinformación y su importancia para las tecnologías de la información geográfica. *Ar@cne. Revista electrónica de re-cursos en Internet sobre Geografía y Ciencias Sociales*. Barcelona: Universidad de Barcelona, nº 148, 1 de junio de 2011. <http://www.ub.es/geocrit/aracne/aracne-148.htm> (02/2013)

Lázaro y Torres, M. L & González González, M.J. 2005. “La utilidad de los Sistemas de Información Geográfica para la enseñanza de la Geografía.” *Didáctica Geográfica* 7: 105-122.

Lázaro, M.L. (coord.) 2012. *Enseñar Geografía a través del análisis de imágenes con la ayuda del campus virtual y la pizarra digital interactiva* (PIMCD 6/2010) Cersa. Available in: <http://eprints.ucm.es/16849/> (02/2013)

—. (coord.) 2012. Webpage with the project in ArcGIS Online embedded: *Learning Geography of Spain on the Web 2.0.* <https://cv2.sim.ucm.es/moodle/course/view.php?id=12502> (02/2013)

Lázaro, M.L.; Alcolea, M.A.; González, M.J. 2013. *La alfabetización digital a través del campus virtual, la Web 2.0 y la Geografía*, UCM. Available in: <http://eprints.ucm.es/20548/> (02/2013)

Lázaro, M.L. (coord.). 2013. “*Aprender Geografía con la Web 2.0*” (Learning Geography with the Web 2.0) (PIMCD 133/2011) (DVD). Available in: <http://www.ucm.es/info/geografiavisor/DVD/> (12/2012)

Milson, A.J. 2011. “SIG en la nube: WebSIG para la enseñanza de la Geografía”. *Didáctica Geográfica,* 11: 111-124.

Milson, A. J.; Demirci, A. & Kerski, J. J. (Eds.). 2012. *International Perspectives on Teaching and Learning with GIS in Secondary Schools*. New York: Springer.

Mínguez García, M.C. 2010. El paisaje como objeto de estudio de la Geografía. Un itinerario didáctico en el marco de la semana de la ciencia de la Comunidad de Madrid. *Didáctica Geográfica,* 11: 37 – 62.

Zwartjes, L. 2012. *Creating a learning line on spatial thinking in education*, Jaar-boek de aardrijkskunde, p. 33□41.

Annex I

Alájar (Huelva, Andalusia) from Peña de Arias Montano

Observation of landscape elements:

In the foreground we see a rural community that has grown around a church. Behind it there is a great forest mass followed by wooded pastureland that follows the *dehesa* model. In the background, blurred by the distance, blue-tinted mountains of a greater altitude than the one on which we are now located.

Analysis and comment:

Above the harmonious settlement rises a church of grand dimensions, with a tower and a pitched roof that covers the central nave. The whole settlement is scattered around this church and spreads from it following the main roads, as is the norm in mediæval villages, following the Strassendorf (street-settlement) layout.

The settlement, situated on a mid-sized mountain, looks well cared for and has no abandoned buildings, which may allow us to assert that there is no population loss, although no growth can be observed either, as there are no tall buildings. The predominant building typology is that of two-storey houses with a white façade and a pitched roof. The bright colour of some of these roofs allows us to infer that they have been repaired recently, which is an indication of a certain moderate economic dynamism. The characteristic white wall, which may be whitewashed, is an indication that this is part of the Andalusian landscape. Abundant rainfall may be deduced from the greenness of the environment and that all roofs are pitched. Thus, we may be in a mid-mountainous area in western Andalusia, which is the mid-mountain area with the heaviest rainfall in Andalusia.

Behind the settlement there appears a forest mass that is varied and irregular, in which the different tree species cannot be distinguished clearly. However, the shape of the trees reveals the absence of pruning. Perhaps this is the shape of chestnut trees (although they cannot be identified in the image). This species is present because of a greater amount of water than is typical of the Mediterranean climate. Cork oaks, holm oaks, pines and even an odd stand of olive trees could also be present. This translates into a varied landscape and a very diversified exploitation thereof. Behind the dense wooded area can be seen an area of mixed pastures and trees following the *dehesa* model, indicating the possible presence of livestock. This latter activity, in view of the area, is especially focused on Iberian pigs and the production of the well-known and delicious ham, *jamón de Jabugo*.

The population seems to be dedicated to exploitation for self-consumption, as there are small to mid-sized plots next to the houses. Because of their size and of their proximity to the river, they could also be small market gardens, although no significant details can be seen because of the distance.

The fact that the whole landscape is hilly indicates that this is a mountain range area. The height of the place from which the photograph was taken is lower than the background area, where settlements are scattered in the middle of the mountain slopes.

Relationship to environmental, economic and social sustainability:

This is an area of great natural and landscape value. The settlement that dominates it is made up of houses of the traditional typology: one or two storeys and a pitched roof, disposed in a labyrinthine lattice around the 16th century San Marcos parish church. All of this is in perfect balance with the environment. This is the Sierra de Aracena and Picos de Aroche area, established as a National Park in July 1989. Alájar, together with Peña de Arias Montano, was declared a historical and artistic complex in 1982, which has been a great boost for a rural tourism intent on enjoying not only the breathtaking landscapes but also natural products, such as those derived from the Iberian pig and from chestnuts.

Internet link: http://en.wikipedia.org/wiki/Al%C3%A1jar

Place and date: Photo of Alájar (Huelva) taken from Arias Montano rock (689 m), a natural monument (April 2010)

Location of the point where the photograph was taken: Time Zone: UTM 29S

- ED50: 37º 52' 43,91" N and 6º 39' 57,47" W

X coordinate: 705282; Y coordinate: 4195012

- WGS84: 37º 52' 39,42" N and 6º 40' 2,33" W

X coordinate: 705156; Y: 4194801

- ETRS89: 37º 52' 39,43" N and 6º 40' 2,37" W

X coordinate: 705155; Y coordinate: 4194803

CHAPTER THIRTEEN

USING THE IBERPIX GEOBROWSER FOR TEACHING GEOGRAPHY: PERSPECTIVES FROM ACTIVE LEARNING METHODOLOGIES

JOSÉ JESÚS DELGADO PEÑA

Introduction

The National Geographic Institute (IGN), created in 1870 under the Spanish Ministry of Development, undertakes a wide range of functions with reference to work in astronomy, geodesy and geophysics. It offers the public—free of charge and in line with the EU INSPIRE Directive—a number of tools that are very useful for understanding the geography of our country from various standpoints and which can be used for educational purposes. The aim of this chapter is to identify and examine some of the IGN products and interesting resources, easily accessible on the IGN website (http://www.ign.es), and more specifically the IBERPIX geo-browser. With the implementation of certain teaching methods, these products can be the basic building blocks for creating useful and effective learning activities in the field of geography and related subject areas. Next, the article presents some interesting methodologies for teaching geography through the use of cartography and subsequently combines both aspects to present some concrete case studies.

IGN's online resources and IBERPIX aerial viewer

The IGN website has a vast number of resources that can be used for educational purposes. This chapter does not attempt to present all of them exhaustively but to highlight what might be considered most relevant to learning and teaching with geo-media, by using IBERPIX as a basis for activities that will later be discussed in terms of active learning methodologies.

Figure 13.1. Detail of the IGN/CNIG home page.

In the central section of its home page (Figure 13.1.) the specific IGN applications are located that are consulted most frequently by users, such as: information on earthquakes, the Cartographic Register, the IBERPIX Viewer, SIANE, Educational Cartography, information on National Parks and the link to the National Astronomical Observatory. Below, there are lots of links associated with IGN, many with an eminently educational focus, such as the case of "My friend the Earth" or "Teaching Atlas of America, Spain and Portugal".

The National Atlas of Spain can be viewed online through the SIANE website, an innovation and development project that gives technological support for production and publication, as well as other products and services offered by the IGN's Department of Thematic Cartography and the National Atlas (Romera, 2012). Website users can perform custom queries of the most current geographic information and access digital content directly (maps, images, data, etc.), since it appears in catalogue format. The enormous potential of this application in the field of education is obvious, as it allows easy access to different kinds of geo-graphical information. This information can then be represented visually, making it much more understandable to students.

Another very interesting application is the SIGNA Geoportal (Spanish National Geo-graphic Information System), which aims to integrate the IGN's digital geographic databases in a GIS (Geographic Information System) for viewing on the Internet and Intranet (Romera, 2012). The interaction with the user is through an interface that allows managing layers of information. This makes it useful for people with little computer knowledge to access and visualise data, and also for experts who need more complex sets of information. Its simplest features include navigating a map, locating addresses or geographical features, measuring lengths and areas, getting information and sharing maps. This resource is ideal for displaying geoinformation in the form of a simple GIS applied to Spain, so that students can gain hands-on understanding of the philosophy and potential of these tools in geography and related disciplines.

The section on the IGN website, Teaching Cartography, is also important. It displays the results of projects whose aim has been to create applications that bring the world of geography and cartography to the classroom. "My friend the Earth", for example, was created for elementary students. Through interactive and entertaining approaches, it favours learning concepts like spatial orientation, maps and globes, the solar system, the lithosphere, hydrosphere, atmosphere and biosphere. The "Atlas of America, Spain and Portugal", aimed at students aged 10 to 14, contributes to their knowledge of the composition and structure of the universe and the Earth, the use of maps and other cartography tools, and especially the geography and history of more than 20 countries in America, Spain and Portugal.

Figure 13.2. Screen menu in IGN's resource, "My friend the Earth"

Another interesting resource is "Spain through maps", created in collaboration with the Spanish Geographic Association (AGE). It offers support materials for ESO (secondary education) and second year high school, and most notably includes the description sections titled

"Landscapes of Spain" and "Population of Spain" that describe the breakdown and distribution of the Spanish population (according to Autonomous Communities, provinces and major cities) through thematic maps and graphics. "Cartographic Concepts" presents visual and easy-to-understand information on the shape and representation of the globe, projections, scale, representation techniques and cartographic symbols, etc. "Your Friend the Map", by F. Aranaz del Río, is a popular work on maps, in its fifth edition. Users can also access and download a variety of puzzles and outline maps of Spain, either for the whole country or else by regions and provinces, Europe and the rest of the world, as well as some online materials related to Secondary Education and designed by the IGN (Geography for ESO and Introduction to Spatial Data Infrastructure).

Another very useful application that is available, and is illustrated in the cases discussed later, is the IBERPIX geo-browser. This is designed to show maps, images and data from the EIONET network, and the National Centre for reference in Land Cover and Uses. It is the system for viewing raster maps, satellite images and orthophotos such as documents with fully valid metric mapping documents (Romera, 2012). Users can make queries to find images based on their metadata and spatial location, viewing and accessing information in detail. The potential for the use of these products in the classroom is significant:

- items can be found by their name or geographic coordinates;
- the topographic map and its SPOT images can be viewed, allowing a side-by-side comparison of the actual image with the map image;
- the zoom function on the orthophoto allows details to be shown, or searching for particular items on a larger scale; comparing existing changes (e.g. in infrastructures or land use) in the National Topographic Map 1:25,000 and the orthophoto can be presented by applying a transparent layer over others (see Figure 13.3.);
- lengths in the orthophoto, such as the distance between two buildings in the neighbourhood can be measured; surfaces in the orthophoto such as a neighbourhood park area can be measured;
- SIOSE terrestrial photos, favouring the understanding of space at ground level and their correspondence to aerial representation, whether as a real image or as a map document, can be viewed;
- maps and orthophotos can be printed to enable working with the print-outs in the classroom, and so on.

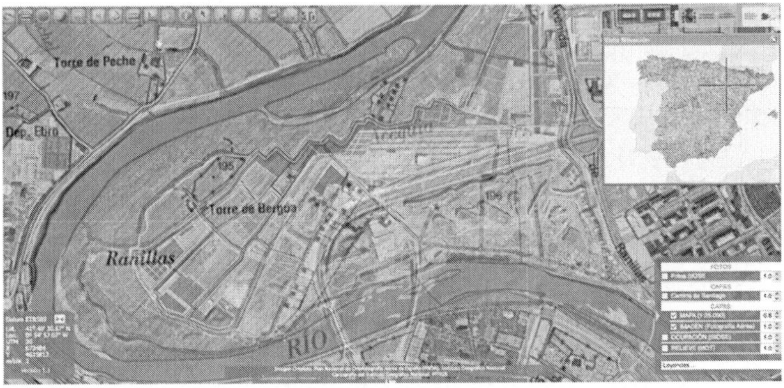

Figure 13.3. IBERPIX screenshot comparing the orthophoto and the topographic 1:25,000 map of the Expo 2008 grounds in Zaragoza (Spain).

In the latest version of IBERPIX, a number of significant improvements have been introduced, including the following, which should be highlighted because of their potential use in teaching geography:

- GPS Routes. Users can download the data from a GPS device, indicating on the map or orthophoto the route taken and the main waypoints. This allows us to develop educational tours with our students or have them develop an itinerary around a specific theme, combining classroom work with fieldwork;
- WMS Viewer. With the use of special glasses, it allows seeing a three-dimensional relief, which greatly facilitates the understanding and comprehension of the territory. Students can relate relief to other geographical aspects such as the location of towns and cities or infrastructure construction.

Finally, there is a lot of map information available from the IGN's Download Centre. It is worth taking a look at all the material available, but mainly the resources that relate to the place where we teach, in order to create a 'local area study' resource bank. Once this is done, as we shall see in the next section, it is important to undertake thorough reflection to develop activities which, on the basis of these materials, will help us in the teaching process, promoting more hands-on and meaningful learning for students, and a more active methodology. Among the resources available here are satellite images, national topographic maps at different scales (1:25,000, 1:50,000, 1:200,000) current and historical, CORINE Project land use maps, Digital Terrain Models with various resolutions, provincial and urban maps (Cartociudad), SIOSE maps (Spanish Land Occupation Information System), etc., in other words, a huge amount of information which would be very useful for teaching and learning geography.

Active teaching methods in combination with cartography: cooperative work, case study method and field work

As already noted, teaching resources require appropriate methodologies as the basis for preparing classroom activities that promote more applied and meaningful geography learning. Cooperative work and case study methodologies are good examples. The new higher education model promoted by the Bologna process reflects a paradigm shift whereby students become the centres and stars of the learning process, thus requiring an active role as opposed to being simple receivers in more traditional teaching environments. Thus, the teacher would go from being a mere transmitter to a true guide, encouraging active learning methodologies and offering learning situations that are more consistent with the professional profile of the studies in question, and where teamwork and the development of personal autonomy focus on "learning how to learn" and become key parts of the process (Delgado, 2010).

In this regard, a major change in the scope of the European Higher Education Area is the growing importance of the role of the learner, who becomes the focal point of the teaching and learning process, in which he or she must play an active role (Bautista, Borges and Forés, 2006. p. 34-36; Benito and Cruz, 2007, p. 15). So, teaching based on skills, abilities and competencies gains predominance, with a consequent increased focus on the professional profile and labour market integration of university graduates (Bautista, Borges and Forés, 2006, p. 22). This system overcomes education and training targeted only on the conceptual and enhances the more open and diversified evaluation processes, both in terms of "learning by doing", which requires an active teaching methodology—most importantly cooperative learning, problem-based learning—and the case method, according to Benito, Bonsón and Icarán, (2007). Achieving greater autonomy in students is vital, so promoting cooperative work is vital (Rué, 2007, p. 154-156).

We strongly believe in the value of the case method approach to accomplishing this. This method was created in the early twentieth century at Harvard University and is currently used throughout the world and in many disciplines. Case studies describe a real event of professional interest, taking into account real-life elements. De Miguel (2006, p. 89) defines it as an "intensive and comprehensive analysis of an event, issue or actual event in order to discover, interpret, solve, generate hypotheses, compare data, reflect, complete knowledge, diagnose and sometimes obtain training in the possible alternative solutions". Thus, "the in-depth analysis of real-life examples dialectically engages students in theory and practice, in a reflective process that becomes, in turn, a meaningful learning process, by having to show or analyse how experts have solved or can solve their problems, the decisions they have taken or could take and the values, techniques and resources involved in each of the possible alternatives". To do this, students need access to necessary information from a range of sources; photos, newspaper clippings, reports, etc., using them to look for the most appropriate solution.

Finally, fieldwork has been traditionally seen as a crucial tool for making learning more interesting and immediate for geography students (Friera, 1995: Hernández, 2007; Liceras, 1997, Souto, 1998). The advantages of field work studies are many, such us direct contact with a certain type of resource, procedurally intensive work, better acquisition of conceptual and attitudinal contents, increased student motivation and addressing problems and real-life scenarios (Hernández, 2007). Moreover, with the development of Information and Communication Technologies, there are devices available, such as GPS, that have increasingly become part of our daily lives and can yield positive authentic results when used in the classroom.

In short, it is vital to always bear in mind the importance of diversification in terms of the methodological strategies in the context of active learning (Rué, 2007), and wherever there are a variety of sources and activities that promote knowledge. When it comes to designing these activities, the valuable resources available from the IGN website can and should be used, as the following examples illustrate. This chapter now presents three cases of the use of the IBERPIX application is used, through which cooperative learning and the case study method, along with fieldwork, take on special relevance.

Some practical activities for cooperative learning and case method application using IBERPIX

Active learning activities need to be developed to promote truly effective teaching and learning processes. It is therefore necessary, based on the resources and tools within our reach, to design attractive, meaningful activities for our students—activities also interesting and educational from an academic standpoint. It's not just a matter of knowledge, but of creativity and originality, of being aware of what is happening around us, and thus, at the right time and with the right resources, to plan an activity that can be used in our classroom. Now, we present some examples that illustrate this activity design process, often linked to chance discoveries in our daily routine.

The case of Torres town centre

Visiting the urban centre of this Jaen town during the Eighth National Congress of Geography Teaching, I became aware of the interesting but tragic story of the street that splits the town in half, and can be seen clearly from the air (Figures 13.4 and 13.5.).

Figure 13.4. Torres town centre in the IBERPIX viewer.

This is a simple case study, based on a real-life example, using real resources (in this case an aerial image). The activity can be carried out in less than an hour, making it ideal for a single classroom session. The aim is to exercise students' powers of observation. Once they have found the image in IBERPIX, they are asked two questions, which they try to answer in teams:

- What layout does the town of Torres (Jaen) have? Briefly describe how it is organised: type of layout, the main orientation of streets, the logical reason for that orientation, etc;
- There is one street in this town, which significantly contrasts with the layout of the other streets. Locate this street, trace it on a map. What might be the origin of this street and why does it have such a

different layout? Look up the necessary information and complementary images online.

Figure 13. 5. Google Street View image showing the Rambla de San Gil in one of the student reports.

The first question should help students realise that most of the streets run parallel and follow the land contours running northwest southeast, a logical configuration to avoid steep inclines. The second question will lead them to observe that there is a single long street running southwest northeast, crossing the town from side to side, dividing it into two equal parts. This street has a steep slope. After proper online research students might discover that this street was caused by a natural disaster, a terrible flood that took place on September 1, 1843, which killed 55 people, in addition to destroying goods and buildings. It is interesting to note that the description of the storm could correspond to the typically Mediterranean and destructive phenomenon known as a “cold drop”. Furthermore, aided by a digital terrain model, also available in the IGN Download Centre, it would be easy to see how the land forms a funnel at the top of the town which leads straight to that ravine, so in the case of heavy rainfall the objects carried by the stream could easily destroy whatever lay in their paths. This explains why the street was created. It is nothing more than an attempt to serve as a drain in case of heavy rainfall. The street was named San Gil, in memory of the date on which the fateful event occurred.

“Malaga Twister”: tornado damage case study

The Malaga region was hit by a tornado on the night of February 1, 2009. Fortunately there were no fatalities, although the economic damage was substantial. This was especially the situation at the *Plaza de la Solidaridad* and in its surrounding areas, where some roofs were completely destroyed, as in the case of the bus station and a nearby car park (Figures 13.6. and 13.7.). In the days that followed, many news reports and images were published, showing Malaga just after the tornado and the progressive damage repair process. Perusing the news in one of the local online newspapers, numerous aerial photographs were available taken from a helicopter showing damages. Many ground-level photographs were also posted. The photographic material was so abundant that I thought of designing the activity shown below, titled “Malaga Twister”.

This project should be carried out in groups and follows the case method approach. To begin with, and as an initial motivation, a document was drafted so students could read about the experiences of a young geographer employed by the City Hall on the night of the tornado. The next day, when he arrives at work, he is given an important task: Draft a report on the damage caused by the tornado in the city of Malaga. Students, grouped in teams, have to help him.

Their task was to use aerial photographs taken just after the disaster and compare them with IBERPIX aerial images in order to perform a visual analysis report for detecting damage and assessing its magnitude. The information given to them at the beginning of the process was as follows:

- A folder with aerial photographs of the area after the tornado and two maps showing the possible path of the tornado through the city;
- A folder with ground-level pictures of the area after the tornado;
- A folder with press reports detailing the tornado incident and the recovery of the area the days following;
- Aerial photography of the west of Malaga (where the tornado hit);
- A sample report (very incomplete) of the work to be done. This report indicated major gaps in red so groups of students could complete it with their own research.

Figure 13.6. Aerial view of Avenida de Europa and adjacent area in the IBERPIX viewer.

The general objective of the study was, first, to determine the type of tornado damage, and secondly, to somehow illustrate the damage through a selection of images. Below is an excerpt of the instructions students had to follow to prepare their reports:

"To prepare the report you have to develop two elements, which are as follows:

- First, the categorisation or classification of the damage caused by the tornado. After viewing the pictures and reading the various news clippings, you should be able to make an organised list of material damage (and easily visible on aerial photographs) to organise a funding programme for persons/entities affected by the disaster. It would be convenient to divide this classification according to whether the damage affects individuals (e.g., a block of flats, etc.), private companies (an industrial warehouse, a business, etc.) or public institutions (a school, a park, street furniture, etc.);
- In addition, based on all the images collected (both aerial and at ground level), you must make a selection of the most significant ones, which best illustrate the list of damages previously established. The minimum number of images chosen will depend on the number of group components. They can be aerial and/or ground-level photos in whatever proportions you consider most appropriate. Both types should be located on the map of west Malaga, available in the resources, with a red box (if it is an aerial photograph, include location and approximate area in the general image), or by a blue circle, if it's a ground-level photograph. In

both cases the location symbol should be accompanied by a tracking number. The reference of this number, together with the image itself, should be included in the report, also showing the following information:

In the case of a ground-level photograph, coloured arrows should indicate the damage that can be seen in the photograph. In the case of an aerial photograph, the areas affected by the path of the tornado should be marked by coloured polygons;

The different colours used should designate the different types of damage. For example, green = vegetation, blue = roofs of private buildings, green = roofs of public buildings, yellow = cars, etc. The work of the team can determine both the classification and the colour mapping to freely designate damage. You can come up with very creative and functional solutions. A location/description of each image needs to be provided, together with a description of the damage. So, good luck and good work!"

The results obtained in this activity were very positive. The groups produced high-quality outcomes and their high levels of motivation and involvement were clear. The main objective of the activity was to use all visual and documentation material found in combination with the IBERPIX application to develop basic geography skills, such as recognition of objects in aerial view shown in the maps, and also recognising objects seen from the air on the ground-level photos. Moreover, by asking them to manipulate images, drawing polygons, arrows, etc., we fostered the acquisition of digital skills. At the same time, when presenting the text, the students must acquire the ability to adequately express geographical features, concepts, and aspects such as location and orientation, thus developing their written skills. All this was done in a workgroup environment, exercising the social skills that are so important in today's professional world.

Figure 13.7. Aerial image of the car park affected.

Setting GPS routes

One of the sessions in the Fifth Seminar on Teaching Geography in Secondary Education in Getafe established the potential of GPS devices in teaching geography (De Lázaro, 2012). Using the new version of the IBERPIX viewer, it is easy to download the route or itinerary that has been saved onto an aerial image or the national topographic map. This allows teachers to design activities in which students come into contact with the terrain to be analysed, learning to apply zoning according to specific features and designing itineraries, which highlight the prominent features of each of the areas being studied.

The following example helps to see this more clearly. A field visit to a nature reserve in our area can be used to supplement the study of various ecosystems or habitats in class. The main features can then be assessed, the dominant plant species, terrain coverage and layers. It was also possible to analyse the area through the use of aerial photos from IBERPIX in order to try to locate these ecosystems and relate them to the terrain visited on the day of the fieldwork visit. Finally, different routes for each working group could be planned in advance, the results then assessed back in the classroom, once they have been downloaded onto computer, and compared to identify their characteristics: for instance, which is the most appropriate, the one that has crossed the most habitats in the least distance.

Another example is the town of Getafe (Figure 13.8.). This city is very interesting to study owing to the historical factors that have shaped it. On the one hand, it is an old town centre, with a significant historical past because of its location mid-way between Madrid, Aranjuez and Toledo. It has a strong commercial and agricultural character, which meant a considerable urban development for its time, culminating in a cathedral built in the late eighteenth century (an impressive and virtually ignored architectural element). On the other hand, it also has a distinctly military air, linked to the air force base, a place recognised as the birthplace of Spanish aviation and where Juan de la Cierva flew his first autogyro (autogiro). This has curtailed, for example, the construction of tall buildings, unlike other cities in the Madrid metropolitan area. A third important factor was its huge industrial development in the early and mid-twentieth century. Finally, the opening the Carlos III University in 1989 has also been decisive in Getafe's development. All these different uses have left their mark on the appearance of the city. Its structure and spatial distribution mirror the various growth stages, patterns that have been moulded to the required functionality at various stages in its history. Following a review and analysis of all these factors in the classroom, a field study visit could be organised in which the various student groups follow different routes where they visit the various "thematic areas" (historical Getafe, the military and aeronautical Getafe, the industrial Getafe and the university Getafe). The final activity would then be for them to determine how the various uses "fit" into the physical and human landscape, resulting in the current city configuration.

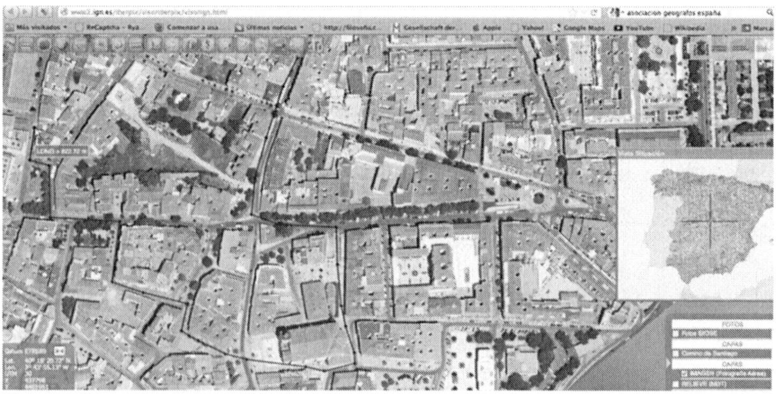

Figure 13.8. Screenshot of IBERPIX with the route taken through Getafe's historic town centre.

Conclusions

In this chapter some case studies and ideas for designing activities have been presented within the framework of active learning, giving special consideration to methodologies considered important under the Bologna Process, such as cooperative learning, the case method and fieldwork. As basic resources to support these activities, applications and resources available on the IGN website, such as IBERPIX, image viewers, maps and data sets can be chosen. Use has also been made of available local information that is relevant to the study and that helps develop activities that are interesting and motivating to students, while at the same time applicable to tasks where they can develop the various skills so necessary today in the labour market.

The case studies demonstrate that if we take advantage of the geographical elements close to our school, college or university, and if we can gather materials that help understand the historical and social aspects that have shaped its urban development, distinctively geographical projects can be carried out in which spatial skills are developed with a practical, meaningful and highly motivating outcomes.

As teachers, we must be aware of developments happening around us, in order to detect possible activities. As has been shown, many things can be useful for creating interesting classroom activities—such as the scars in an urban centre from an ancient catastrophe, a tornado's path, or a simple view of the structural complexity of any city. These are exercises that require creativity and originality to develop, where effort is needed to link applications and resources with teaching methodologies and techniques that contribute towards meaningful and long-lasting learning. Data is easily forgotten and can always be located from the right sources, but skills or competencies remain part of our learning much longer. This is much more the case when they have been developed working in close proximity, with a hands-on approach, and working side-by-side with people who have joined us in our training process, both peers and teachers.

References

Bautista, G., Borges, F. & Forés, A. 2006. *Didáctica universitaria en Entornos Virtuales*. Madrid: Narcea.

Benito, A., Bonson, M. & Icarán, E. 2007. "Metodologías activas" in A. Benito and A. Cruz. *Nuevas claves para la Docencia Universitaria en el EEES*. Madrid: Narcea.

Benito, A. & Cruz, A. 2007. *Nuevas claves para la Docencia Universitaria en el EEES.* Madrid: Narcea.

De Lázaro y Torres, M. L. 2012 "El trabajo con GPS: funcionamiento, salida de campo y tratamiento de la información". *V Seminario sobre la enseñanza de la Geografía en la Educación Secundaria*. Getafe: unpublished.

De Miguel Díaz, M. 2006. *Modalidades de enseñanza centradas en el desarrollo de competencias*. Oviedo University of Oviedo.

Delgado Peña, J. J. 2010. "Metodologías activas de enseñanza-aprendizaje y evaluación en el ámbito del E-learning. Una experiencia desde la asignatura Cartografía Fundamental" in *Geography, Teacher Education in the European Area of Higher Education*. Madrid: Grupo de Didáctica de la AGE, p. 237-252.

Friera, F., 1995. *Didáctica de las Ciencias Sociales. Geografía e Historia.* Madrid: Ediciones de la Torre.

Hernández, F. X., 2007. *Didáctica de las ciencias sociales, geografía e historia*, 3rd ed. Barcelona: Graó.

Liceras, A., 1997. " La observación y el estudio del paisaje" in L. García, coord. *Didáctica de las Ciencias Sociales, Geografía e Historia en la Enseñanza Secundaria*. Granada: Grupo Editorial Universitario, p. 297-325.

Romera Sáez, C. 2012. "La información geográfica en el IGN/CNIG (Web)". *V Seminario sobre la enseñanza de la Geografía en la Educación Secundaria*. Getafe: unpublished.

Rue, J. 2007. *Enseñar en la Universidad. El EEES como reto para la Educación Superior*. Madrid: Narcea.

Souto, X. M., 1998. *Didáctica de la Geografía. Problemas sociales y conocimiento del medio*. Barcelona: Ediciones del Serbal.

CONTRIBUTORS

Editors

Rafael de Miguel González. PhD in urban studies and planning at the *Institut d'Urbanisme de Paris*, University of Paris. Associate professor of geographical education and teacher training at the University of Zaragoza, Faculty of Education, where he is Vice-Dean for International Relations. He has also been a teacher in secondary education. Member of GEOT (Research group on spatial planning). Director of a research project in spatial planning for the Spanish Government and member of research projects on urban geography and geographical education. He belongs to the Spanish Geographers Association and to the Royal Geographic Society of Spain (both partners of digital-earth.eu, the second one as a Centre of Excellence). He is Fellow member of the Royal Academy of San Luis. More than 40 publications (papers in JCR journals, books, chapters book) in the research areas: geographical education (innovative teaching and learning geography with Geographical Information Technologies); urban geography; urban and regional planning. rafaelmg@unizar.es

Karl Donert. He is a consultant and education specialist. He is Director of the European Centre of Excellence, digital-earth.eu, based at Salzburg University, Austria. He is a UK National Teaching Fellow and worked at Liverpool Hope University. He was coordinator of the HERODOT Network for Geography in higher education and is President of EUROGEO (the European Association of Geographers). He is a member of the Council of Europe Education Group, an expert for the European Commission in Brussels and former Hon. Vice President of the Geographical Association. He researches and publishes widely on many aspects of geographical education, citizenship and on the State of Geography in Europe. kdonert@yahoo.com

Contributors

Eyüp Artvinli has been an associate professor in the Faculty of Education at the Eskisehir Osmangazi University and is also the current RIGEO Editor-In-Chief at www.rigeo.org. RIGEO (Review of International

Geographical Education Online). His main focus areas are curricula in high schools, textbooks, active learning, teacher training, digital geography education, and GIS in schools in geography. Dr. Artvinli is also founder of a Turkish Geography Teachers Association.

Isaac Buzo. Geography teacher in the "San Roque" secondary school in Badajoz (Spain). A member of the Association of Spanish Geographers, as well as being on board. He has developed educational resources for various publishers. He has his own educational website on social science resources (www.isaacbuzo.com).

Livio Chiarullo. PhD in "Planning and Public Policies" at the University of Venice, degree in Political Sciences and a Master's in Eco-Management. He has worked with public and private agencies for definition and implementation of bottom up sustainability policies. He currently works as a Senior Researcher at the Fondazione Eni Enrico Mattei (FEEM).

Giovanni Donadelli. PhD student at the University of Padua, Giovanni is an expert on educational technologies. His PhD research focus is on participatory approaches within geographical education and on teaching strategies for teaching geography using technologies.

Jesús Delgado Peña. Associated professor in Málaga University. He has conducted field-research on physical geography and biogeography for 16 years. He has been especially occupied with questions relating to vegetation dynamics, environmental planning and desertification processes monitoring mainly in a Mediterranean context. He has recently also been interested in pedagogic methodologies in relation to new information technologies and geography with emphasis on elderly students.

Tim Favier is Post–Doc. researcher in the field of geography education at the Centre for Educational Training, Assessment and Research of VU University Amsterdam in the Netherlands, and also geography teacher trainer at Fontys Hogeschool Tilburg in the Netherlands. He is a specialist in Geographic Information Systems (GIS) and wrote his PhD study on the use of GIS in secondary education.

María Jesús González González. Ph.D. in Geography. Associated professor in León University. Member of the Board of the Geographical Education group in Spanish Geographers Association. Advisory board of

Didáctica Geográfica Journal. Member of the Real Sociedad Geográfica of Spain. Member of the research group of the Complutense University of Madrid (UCM 931335-GEODIDAC) dealing with teaching & learning geography in higher education. She has written papers and articles on GIT in geographical education.

Lea Houtsonen is Counsellor of Education at the Finnish National Board of Education. Member of the Finnish Examination Board (Leader of Geography Section). She is a Docent in the University of Helsinki and former Chair of the Commission on Geographical Education of the International Geographical Union.

Aikaterini Klonari is an Assistant Professor in Teaching Geography at the University of the Ægean. Her research focus is on Methods in Teaching and Learning, Curriculum Development, Use of ICT and GIS in Education—particularly in teaching geography. She is coordinator and co-author of the National Geography Curriculum, geography textbooks and teachers' guides. She is a member of IGU, AAG, EUROGEO, ATEE, ICET, GGU.

Maria Luisa de Lázaro is Assistant Professor at University Complutense of Madrid. She has been President of Geographical Education group in the Spanish Geographers Association and editor of the journal *Didactica Geografica*. Member of the board of the Royal Geographic Society of Spain and Director of the Spanish Centre of Excellence digital-earth.eu/Real Sociedad Geográfica. Head of the research group of the Complutense University of Madrid (GEODIDAC) dealing with teaching & learning geography. She has written text books, papers and articles on GIT in geographical education.

Sanna Mäki is Senior Lecturer in geography at University of Turku. President of the Geographical Society of Finland. Special research interests are related to regional geography, environmental planning, geography education and geo-media, etc.

Cristiana Martinha is a Post–Doc. Fellow in the University of Porto and a researcher at CEGOT (Center of Studies in Geography and Spatial Planning). She is also visting assistant professor at University of Minho. She was fellow and internee at Georg-Eckert-Institut in Germany. She has published on geographical education.

Carmen Mínguez is assistant professor of geography at University Complutense of Madrid. She has teaching GIS in an Archaeological degree. She has written textbooks, papers and articles on cultural tourism, historical cities and geographical education.

Piero Morseletto is a PhD researcher at the Institute for Environmental Studies (IVM), VU University Amsterdam. His PhD research examines the political side of the science-policy interface, specifically how the scientific environmental knowledge is absorbed and interpreted by decision-makers.

Juna Papajorgji, PhD in Urban Planning, is an Adjunct Faculty at the Department of Urban and Regional Planning and a Research Associate at the GeoPlan Centre, at the University of Florida. Her interests include GIS, and Open Technologies for open societies. She is the co-founder of the GISCorps, a former board director of the Urban and Regional Information Systems (URISA) association, and the current Vice Chair of the Societal Impacts Committee of the Global Spatial Data Infrastructure (GSDI) association.

Maria Pigaki is Teaching Associate at National Technical University of Athens. She has researched into geodesy, cartography and geography projects using GIS. Her book entitled "Enseigner la Cartographie" was published this year by the "Editions Universitaires Europeennes" (EUE)-Denmark.

Juha Riihelä is PhD candidate and project researcher at the University of Turku, Department of Geography.

Lorena Rocca is researcher at the University of Padova, where she teaches geography and didactics of geography. She was part of the FEEM research team, and Co-ordinator of the PANDORA project.

Joop van der Schee is professor of geography education and communication at Utrecht University and special professor at the Centre for Educational Training, Assessment and Research of VU University Amsterdam. He is co-chair of the Commission on Geographical Education of the International Geographical Union.

Tuuli Toivonen is a researcher and a senior lecturer in geo-informatics at the University of Helsinki. She has a long time interest in applying GIS in geography education at different educational levels.

Juka Tulivuori has a Master's in education and geography. He is currently working in the Finnish National Board of Education as a Counsellor of Education. His main areas are the use of ICT in education and learning and curriculum development.

Luc Zwartjes is Fellow in Geography and works as teaching assistant & researcher in the Geography Department of Ghent University. He also teaches geography part-time at a secondary school and is co-author of several geography school books. Luc is chairman of the Flemish Geography Teachers Association (VLA) and member of the board of the European Association of Geographers (EUROGEO). He is Director of the Belgian Centre of Excellence digital-earth.eu. He is also linked to the iGeo Task force of the IGU, responsible for the yearly international geo-Olympiad.